Modern Methods of Plant Analysis

Volume 15

Editors
H. F. Linskens, Nijmegen/Siena/Amherst
J. F. Jackson, Adelaide

Alkaloids

Edited by
H. F. Linskens and J. F. Jackson

Contributors

R. A. Andersen   J. C. Callaway   R. C. Crouch   D. Dagnino
M. M. Gupta   J. D. Hamill   G. E. Martin   T. Naaranlahti
P. R. Nelson   M. W. Ogden   A. J. Parr   J. Schripsema
J. R. Sharma   R. Verpoorte   W. W. Weeks   C. A. Wilkinson

With 58 Figures

Springer-Verlag
Berlin Heidelberg New York
London Paris Tokyo
Hong Kong Barcelona
Budapest

Prof. Dr. HANS FERDINAND LINSKENS
Goldberglein 7
D-91056 Erlangen, Germany

Prof. Dr. JOHN F. JACKSON
Department of Horticulture, Viticulture and Oenology
Waite Agricultural Research Institute
University of Adelaide
Glen Osmond, S.A. 5064
Australia

ISBN-13:978-3-642-84228-3   e-ISBN-13:978-3-642-84226-9
DOI: 10.1007/978-3-642-84226-9

The Library of Congress Card Number 87-659239 (ISSN 0077-0183)

Production Editor: Herta Böning, Heidelberg
Typesetting: Best-set, Hong Kong
31/3130-5 4 3 2 1 0 – Printed on acid-free paper

Introduction

Modern Methods of Plant Analysis

When the handbook *Modern Methods of Plant Analysis* was first introduced in 1954 the considerations were:
1. the dependence of scientific progress in biology on the improvement of existing and the introduction of new methods;
2. the difficulty in finding many new analytical methods in specialized journals which are normally not accessible to experimental plant biologists;
3. the fact that in the methods sections of papers the description of methods is frequently so compact, or even sometimes so incomplete that it is difficult to reproduce experiments.
These considerations still stand today.

The series was highly successful, seven volumes appearing between 1956 and 1964. Since there is still today a demand for the old series, the publisher has decided to resume publication of *Modern Methods of Plant Analysis*. It is hoped that the New Series will be just as acceptable to those working in plant sciences and related fields as the early volumes undoubtedly were. It is difficult to single out the major reasons for success of any publication, but we believe that the methods published in the first series were up-to-date at the time and presented in a way that made description, as applied to plant material, complete in itself with little need to consult other publications.

Contribution authors have attempted to follow these guidelines in this New Series of volumes.

Editorial

The earlier series *Modern Methods of Plant Analysis* was initiated by Michel V. Tracey, at that time in Rothamsted, later in Sydney, and by the late Karl Paech (1910–1955), at that time at Tübingen. The New Series will be edited by Paech's successor H. F. Linskens (Nijmegen, The Netherlands) and John F. Jackson (Adelaide, South Australia). As were the earlier editors, we are convinced "that there is a real need for a collection of reliable up-to-date methods for plant analysis in large areas of applied biology ranging from agriculture and horticultural experiment stations to pharmaceutical and technical institutes concerned with raw material of plant origin".

The recent developments in the fields of plant biotechnology and genetic engineering make it even more important for workers in the plant sciences to become acquainted with the more sophisticated methods, which sometimes come from biochemistry and biophysics, but which also have been developed in commercial firms, space science laboratories, non-university research institutes, and medical establishments.

Concept of the New Series

Many methods described in the biochemical, biophysical, and medical literature cannot be applied directly to plant material because of the special cell structure, surrounded by a tough cell wall, and the general lack of knowledge of the specific behavior of plant raw material during extraction procedures. Therefore all authors of this New Series have been chosen because of their special experience with handling plant material, resulting in the adaptation of methods to problems of plant metabolism.

Nevertheless, each particular material from a plant species may require some modification of described methods and usual techniques. The methods are described critically, with hints as to their limitations. In general it will be possible to adapt the methods described to the specific needs of the users of this series, but nevertheless references have been made to the original papers and authors. While the editors have worked to plan in this New Series and made efforts to ensure that the aims and general layout of the contributions are within the general guidelines indicated above, we have tried not to interfere too much with the personal style of each author.

There are several ways of classifying the methods used in modern plant analysis. The first is according to the technological and instrumental progress made over recent years. These aspects were used for the first five volumes in this series describing methods in a systematic way according to the basic principles of the methods.

A second classification is according to the plant material that has to undergo analysis. The specific application of the analytical method is determined by the special anatomical, physiological, and biochemical properties of the raw material and the technology used in processing. This classification was used in Volumes 6 to 8, and for some later volumes in the series.

A third way of arranging a description of methods is according to the classes of substances present in the plant material and the subject of analytic methods. The latter will be used for later volumes of the series, which will describe modern analytical methods for alkaloids, drugs, hormones, etc.

Naturally, these three approaches to developments in analytical techniques for plant materials cannot exclude some small overlap and repetition; but careful selection of the authors of individual chapters, according to their expertise and experience with the specific methodological technique, the group of substances to be analyzed, or the plant material which is the subject of chemical and physical analysis, guarantees that recent developments in analytical methodology are described in an optimal way.

Volume Fifteen – Alkaloids

The new series in *Modern Methods of Plant Analysis* would not be complete without dealing with the alkaloids. These natural products have been the subject of study and practical application medicinally since early in the 19th century. It is timely, then, that here in this 15th volume of the series we have gathered together a collection of chapters on the latest methods used in the analysis of this important group of compounds. Since the 1950s, analysis has been radically changed, to include the various powerful spectroscopic methods which inevitably dominate this volume.

Volume fifteen begins with an "introductory" chapter on isolation, identification, and structure analysis of the alkaloids, identifying the various strategies involved in dealing with these compounds. Then follows a chapter devoted to one of the latest and most powerful of the spectroscopic methods, that of inverse-detected 2D NMR, and its application to alkaloid chemistry. Immediately following this is a chapter on analysis of alkaloids by liquid chromatography coupled with mass spectrometry. The problems associated with the interface between the liquid chromatography column and the mass spectrometer are outlined, including a discussion of the thermospray, electrospray, ionspray, and heated nebulizer interfaces. It is pointed out that in the alkaloid analysis field, liquid chromatography coupled to mass spectrometry is not in routine use. A major drawback is that the commercially available systems utilize only soft ionization methods, which yield very little fragmentation and thus no possibility of distinguishing between isomers, for example. An interesting chapter then follows, dealing with liquid chromatography involving electrochemical detection of the alkaloids. This method is quite sensitive and should find greater use in the future in conjunction with HPLC. Gas chromatography as coupled with mass spectrometry (GC/MS) is the subject of another chapter, and deals in detail with aspects of this coupling.

A consideration of the various aspects of alkaloids in tobacco is the subject of several chapters. Firstly, consideration is given to the production of alkaloids in flue-cured tobacco; this is followed by a chapter dealing with the assessment of burley and dark tobacco alkaloids during storage, aging, and fermentation. Changes that take place during these stages are pointed out. Since there has been much discussion recently on the "passive" effects of tobacco smoke, a chapter on the analysis of "mainstream" and "sidestream" or passive tobacco smoke is both relevant and timely. The complexity of sampling and analysis is quite apparent in this interesting chapter. A further chapter follows on the major analytical methods for the determination of tobacco-specific N-nitrosamines and areca-derived N-nitrosamines, both arising from alkaloids, in tobacco smoke and in the betel quid.

The volume is rounded of by a further two chapters, one on the methods for production of alkaloids in root cultures and their analysis. These root cultures can be established from alkaloid synthesizing dicotyledonous plants. The final chapter deals with genetic and chemical analysis for alkaloids in *Papaver*, a plant of considerable medicinal significance.

Acknowledgments. The editors express their thanks to all contributors for their efforts in keeping to production schedules, and to Dr. Dieter Czeschlik and the staff of Springer-Verlag for their cooperation in preparing this and other volumes in the series *Modern Methods of Plant Analysis.*

Nijmegen/Siena and Adelaide, 1994   H. F. LINSKENS
                                                              J. F. JACKSON

Contents

**Isolation, Identification, and Structure Elucidation of Alkaloids
A General Overview**
R. Verpoorte and J. Schripsema

Inverse-Detected 2D-NMR Applications in Alkaloid Chemistry
G. E. Martin and R. C. Crouch

Electrochemical Detection of Alkaloids in HPLC
V.-P. RANTA, J. C. CALLAWAY, and T. NARRANLAHTI

Gas Chromatography in the Analysis of Alkaloids
D. DAGNINO and R. VERPOORTE

Alkaloid Analysis in Flue-Cured Tobacco
C. A. WILKINSON and W. W. WEEKS

**Assessment of Burley and Dark Tobacco Alkaloids During Storage,
Aging, and Fermentation**
R. A. ANDERSEN

Detection of Alkaloids in Environmental Tobacco Smoke
M. W. OGDEN and P. R. NELSON

**Methods for Production of Alkaloids in Root Cultures
and Analysis of Products**
J. D. HAMILL and A. J. PARR

Genetic and Chemical Analysis for Alkaloids in *Papaver*
J. R. SHARMA and M. M. GUPTA

Contents   XV

List of Contributors

ANDERSEN, ROGER A., United States Department of Agriculture, Agricultural Research Service, Department of Agronomy, University of Kentucky, Lexington, KY 40546-0091, USA

CALLAWAY, J.C., Department of Pharmaceutical Chemistry, University of Kuopio, P.O. Box 1627, FIN-70211 Kuopio, Finland

CROUCH, RONALD C., Division of Organic Chemistry, Burroughs Wellcome Co., 3030 Cornwallis Rd., Research Triangle Park, NC 27709, USA

DAGNINO, D., Division of Pharmacognosy, Center for Bio-Pharmaceutical Sciences, Leiden University, P.O. Box 9502, N-2300 RA Leiden, The Netherlands

GUPTA, M.M., Department of Phytochemical Technology, Central Institute of Medicinal and Aromatic Plants (CIMAP), P.O.-CIMAP, Lucknow-226015, India

HAMILL, JOHN D., Department of Genetics and Developmental Biology, Monash University, Clayton 3168, Melbourne, Victoria, Australia

MARTIN, GARY E., Division of Organic Chemistry, Burroughs Wellcome Co., 3030 Cornwallis Rd., Research Triangle Park, NC 27709, USA

NAARANLAHTI, TOIVO, Department of Pharmaceutical Chemistry, University of Kuopio, P.O. Box 1627, FIN-70211 Kuopio, Finland

NELSON, PAUL REDFIELD, III, R.J. Reynolds Tobacco Company, Research & Development, Winston-Salem, NC 27102, USA

OGDEN, MICHAEL WAYNE, R.J. Reynolds Tobacco Company, Research & Development, Winston-Salem, NC 27102, USA

PARR, ADRIAN J., AFRC Institute of Food Research, Norwich Laboratory, Colney, Norwich NR4 7UA, United Kingdom

RANTA, VELI-PEKKA, Department of Pharmaceutical Chemistry, University of Kuopio, P.O. Box 1627, FIN-70211 Kuopio, Finland

SCHRIPSEMA, JAN, Division of Pharmacognosy, Leiden/Amsterdam Center for Drug Research, Leiden University, P.O. Box 9502, N-2300 RA Leiden, The Netherlands

SHARMA, J. R., Central Institute of Medicinal and Aromatic Plants (CIMAP), P.O.-CIMAP, Lucknow-226015, India

VERPOORTE, ROBERT, Division of Pharmacognosy, Leiden/Amsterdam Center for Drug Research, Leiden University, P.O. Box 9502, N-2300 RA Leiden, The Netherlands

WEEKS, WILLARD W., Department of Crop Science, North Carolina State University, Raleigh, NC 27695, USA

WILKINSON, CAROL A., Virginia Tech Southern Piedmont, Agricultural Research and Extension Center, P.O. Box 448, Blackstone, VA 23824, USA

Isolation, Identification, and Structure Elucidation of Alkaloids
A General Overview

R. Verpoorte and J. Schripsema

1 Introduction

The first alkaloids were already isolated in the early 19th century (e.g., morphine, strychnine). Although the methods for identification and structure elucidation have changed a great deal, the methods of isolation used in the last century are still widely used. Originally, pure chemistry, like derivatization and degradation, was used to unravel the often complex structures of alkaloids. The structure elucidation of a well-known alkaloid, such as, for example, strychnine, took almost 140 years after its first isolation by Pelletier and Caventou in 1818.

In the past 40 years, structure elucidation has been revolutionized by the introduction of various spectrometric methods. In the 1950s, ultraviolet (UV) and infrared (IR) spectroscopy developed into major tools for structure elucidation. In the 1960s, mass spectrometry (MS) and proton nuclear magnetic resonance (^{1}H-NMR) spectrometry (60–100 MHz) had a major impact on the strategies for structure elucidation. In the 1970s, carbon-13 nuclear magnetic resonance (^{13}C-NMR) spectrometry became available, and in the last decade high resolution FT-NMR, including two-dimensional methods. Both methods again changed the world of the phytochemist.

Besides the spectrometric methods, chromatography has developed as a major tool in the isolation and identification of natural products. Present methods allow the rapid separation of small amounts of compounds, and the detection limits for natural products in plants are still further pushed to lower levels.

Here, we will discuss the general strategies followed in the isolation, identification, and structure elucidation of alkaloids. The various methods will be briefly discussed separately, with special reference to the applications for alkaloid analysis. We will not try to give a complete review of all work done on the structure elucidation of alkaloids. For reviews on the different classes of alkaloids we refer to the series *The Alkaloids*, Volumes 1–5 edited by Manske and Holmes (1950–1955), Volumes 6–16 by Manske (1955–1977), Volumes 17–20 by Manske and Rodrigo (1979–1981) and later volumes by Brossi (1983–1992), and *Alkaloids: Chemical and Biological Perspectives* edited by Pelletier (since 1983). Furthermore, numerous review articles on different groups of alkaloids have been published; particularly on isoquinoline alkaloids a series of reviews has appeared in the Journal of

Modern Methods of Plant Analysis, Volume 15
Alkaloids (ed. by Linskens/Jackson)
© Springer-Verlag Berlin Heidelberg 1994

Natural Products. However, it would be beyond the scope of this chapter to give the references to all these publications. For a quite comprehensive collection of data on alkaloids the reader is referred to the *Encyclopedia of Alkaloids* (Glasby 1975) and to the more recent *Dictionary of Alkaloids* (Southon and Buckingham 1989).

2 Isolation

2.1 Extraction

Most alkaloids are colorless compounds; only a few highly conjugated compounds are colored (e.g., berberine, serpentine) or show strong fluorescence (e.g., quinine). Many alkaloids are difficult to crystallize as a free base, but do crystallize as a salt.

Almost all alkaloids have basic properties. The pK_a values vary from about 6–12, with most alkaloids in the range of 7–9. In general, the free base is soluble in organic solvents and not in water. Protonation of the tertiary nitrogen in the free base usually results in a water-soluble compound. This characteristic is used in the selective isolation of alkaloids. Quaternary alkaloids are poorly soluble in organic solvents but are soluble in water at any pH.

The methods for the isolation of alkaloids are based upon the fact that they can be extracted under neutral or basic conditions (after basification of, e.g., the plant material or biofluid to pH 7–9 with ammonia, sodium carbonate, or sodium bicarbonate), as free base with organic solvents (e.g., dichloromethane, chloroform, ethers, ethyl acetate, alcohols) and as protonated base with polar solvents (water, alcohols) under acidic conditions (after acidification to pH 2–4 with diluted acids like phosphoric acid, sulfuric acid, citric acid). Some alkaloids can only be extracted at higher pH (>10), e.g., tryptamine. On the other hand, alkaloids containing phenolic groups dissociate at higher pH, and are thus not extracted by organic solvents under such conditions (e.g., morphine). Further purification can be done by liquid/liquid extraction or liquid/solid extraction.

In the classical liquid/liquid extraction methods, the alkaloids are, after basification, extracted from an aqueous solution with an immiscible organic solvent (e.g., dichloromethane, diethyl ether, ethyl acetate, chloroform); or from an organic solvent with a diluted aqueous acid solution (e.g., phosphoric acid, sulfuric acid, citric acid).

With the aid of ion-pairing agents (e.g., alkylsulfonic acids), alkaloids, including quaternary nitrogen compounds, can be extracted from an acidic aqueous solution with organic solvents. It should be noted that common anions as Cl^-, Br^-, I^- and acetate also result in ion pairs readily soluble in organic solvents. In liquid/liquid extractions this may result in poor recoveries. To avoid this, acids like phosphoric acid, sulfuric acid, and citric acid should be preferred in such procedures (Hermans-Lokkerbol and Verpoorte 1986).

In liquid/solid extractions alkaloids are bound to a solid matrix. Reversed phase materials such as chemically bonded C_8 and C_{18} on silica are widely used. The alkaloids are concentrated as free base on these columns, e.g., from aqueous neutralized extracts, and subsequently washed from the column by a suitable (usually partly organic) eluent. Also ion-exchange materials are used for the selective extraction of alkaloids (for a review, see Verpoorte and Baerheim Svendsen 1984; Popl et al. 1990).

For preparative purposes, purifications based on the precipitation of alkaloids are sometimes employed. A crude extract of the alkaloids is made with aqueous acid, subsequently the alkaloids are precipitated with reagents such as Mayer's reagent (1 M mercury chloride in potassium iodide) or Reinecke's salt (5% ammonium reineckate in 30% acetic acid) at pH 2, or picric acid (saturated aqueous solution) at pH 5–6. After collection of the precipitate by filtration or centrifugation, the precipitate is dissolved in an organic solvent (acetone-methanol-water 6:2:1). The complexing group is then removed by means of an anion exchanger (Jordan and Scheuer 1965; Verpoorte and Baerheim Svendsen 1976). This method is particularly suited for the purification of quaternary alkaloids.

2.2 Chromatography

For the large-scale separation of alkaloids, column chromatography is widely used, silica gel and aluminum oxide being the stationary phases of choice. Due to the acidic properties of silica gel, severe tailing of the basic alkaloids may occur. Addition of small amounts of basic compounds (e.g., ammonia) to the eluent, as in thin-layer chromatography (TLC) systems, will improve the separation. Reversed phase stationary phases can also be used for preparative purposes, but due to the high costs of these materials, they are usually applied in the last steps of a purification when smaller columns can be used. Also with reversed phase materials, severe tailing may occur due to residual acidic silanol groups. For analytical separations the addition of small amounts of (long chain) amines (e.g., hexylamine) or triethylamine may reduce tailing. Furthermore, special reversed phase stationary phases have been developed for basic compounds.

For preparative purposes counter-current chromatography (CCC) has gained renewed interest in recent years (Conway 1990). Efficient and fast separations are possible with newly developed techniques such as droplet counter-current chromatography, centrifugal partitioning counter-current chromatography and coil centrifugal counter-current chromatography. For such CCC separations, ion-pair gradients have been described as selective means of separation of alkaloids (Hermans-Lokkerbol and Verpoorte 1986; Van der Heijden et al. 1987).

2.3 Artifacts

Alkaloids are often rather unstable, e.g., N-oxidation is quite common. In addition to by heat and light, the stability is influenced by solvents (for a review see Baerheim Svendsen and Verpoorte 1983). Halogen-containing solvents are widely used in alkaloid research, and chloroform in particular is one of the most suitable solvents, because of its relatively strong proton donor character. However, these solvents are very active in terms of artifact formation. In chloroform (N-)oxidations occur readily. Also peroxides in ethers may rapidly cause N-oxidations. With dichloromethane, quaternary N-dichlorometho compounds may be formed (Phillipson and Bisset 1972). Similar compounds are formed with minor impurities present in chloroform. Moreover, in chloroform, phosgene is formed, which reacts with the stabilizer ethanol, yielding ethyl chloroformate. This compound may react with secondary amines, causing the formation of ethylcarbamates (Siek et al. 1977).

Particularly in the analysis of trace amounts of alkaloids, e.g., in studies of metabolism, such minor impurities in solvents may have great influence.

Generally, alkaloids are more stable in toluene, ethyl acetate, and alcoholic solutions. In the case of alkaloids containing carbinolamine functions, reactions with alcohols (chloroform contains 1–2% of ethanol as stabilizer!) will occur (e.g., O-methyl pseudostrychnine formed from pseudostrychnine (Bisset et al. 1965). Such carbinolamines, among others, are often found as oxidation products formed from N-oxides or as intermediates in biosynthetic pathways.

Ketones such as acetone and methylethylketone are well known artifact formers. Berberine, for example, may react with acetone (Beke 1963). Ammonia in combination with acetone may react during column chromatography, yielding condensates that give a Dragendorff-positive reaction. Ammonia may also react with aldehydes present in plant materials, giving rise to artificial alkaloids, e.g., gentianine, which is formed from sweroside during extraction.

2.4 Selective Detection

Based on the isolation procedure followed, one can conclude with some degree of certainty whether an isolated compound is an alkaloid or not. Particularly acid/base extraction schemes result in rather specific extraction of alkaloids. Moreover, several reagents that react specifically with alkaloids have been described, e.g., Dragendorff's reagent, potassium iodoplatinate, Mayer's reagent. The first two reagents mentioned are also used as detection agents in TLC. It must be kept in mind that both reagents also react with some nonalkaloidal compounds (Anderson et al. 1977; Baerheim Svendsen and Verpoorte 1983; Verpoorte et al. 1983). In the identification of alkaloids, TLC in combination with selective color reagents is a very useful method. For several classes of indole alkaloids, like *Catharanthus* (Farnsworth et al. 1964) and *Tabernaemontana* alkaloids (Van Beek et al. 1984), schemes for the identification have been described using such an approach. Several of these color reactions give information on certain structural elements in the alkaloids and can thus be used in structure elucidation of unknown compounds.

3 Identification and Structure Elucidation

The identification of an alkaloid is first of all a matter of classification. The source from which the compound is isolated will give information on its nature. If it is from a biofluid, e.g., urine or blood, then most likely the compound is a drug, dope, drug of abuse, or metabolite of such a compound. If it is from plant material, chemotaxonomy will provide information about the type of alkaloid commonly found in that particular plant, plant genus, or family. For example, if one has isolated a basic compound from a plant belonging to the family Menispermaceae, it is most likely an isoquinoline alkaloid, a class of which already about 4000 compounds have been described. On the other hand, plants from the families Apocynaceae, Loganiaceae, or Rubiaceae are most likely to contain indole alkaloids, of which about 4100 compounds are known. Knowing the genus to which a plant belongs is a further way to reduce the number of possibilities. In the three indole alkaloid-containing families mentioned, in each of the most important alkaloid-bearing genera, about 200–300 different alkaloids have been found, some of which are typical for the genus, others being more or less ubiquitous for these families.

In the case of known alkaloids, a complete isolation is not necessary. In fact an alkaloid can be identified with great certainty by co-TLC, co-GLC, and/or co-HPLC using several different separation systems. Also a mass spectrum (e.g., obtained after GC-MS of a crude extract) may be sufficient to identify a known compound; however, closely related alkaloids can be difficult to distinguish by a mass spectrum only.

Known alkaloids can be detected in complex mixtures also by NMR. ^{1}H-NMR has been used for this purpose (Schripsema and Verpoorte 1991). ^{13}C-NMR is even better suited because of the large shift range, which enables the observation of all signals with a sufficient signal-to-noise ratio, also when impurities are present. By comparison with reported spectra, the presence of a known compound can be established with great certainty.

However, the drawback of this technique is that relatively large amounts of compound are needed to obtain a good spectrum. For the analysis of alkaloids, this approach has not yet been used. For essential oils it has been extensively described by Formacek and Kubeczka (1982).

For identification of known alkaloids in crude mixtures 2D-NMR seems very suitable, especially 2D-COSY, which combines a good sensitivity with a better resolution.

In the case of unknown alkaloids, the identification can be accomplished easily if the alkaloid is a simple derivative of a known alkaloid. Such a derivative of a known alkaloid might be recognized by comparing UV and MS data. One can think of simple derivatives having, for example, extra hydroxy (M+16), methoxy (M+30), acetyl (M+42), or N-oxide (M+16) substituents. Such derivatives will easily be recognized in the mass spectrum because the molecular ion and some of the fragments will be shifted with these mass numbers. However, it has to be kept in mind that MS is a destructive method. Although only very small amounts are needed for MS (microgram range), in the case of only very small amounts of alkaloid being available, nondestructive spectrometric methods should be run first.

In the case of none of the above-mentioned methods resulting in an identification or a proposal for a structure, a really novel structure might be involved. In that case, further spectral data will be needed (^{1}H-NMR, ^{13}C-NMR, IR) from which various structural elements may be deduced. Eventually, these results can be combined with the knowledge of the biosynthetic pathways for the class of alkaloids concerned, and possible structures can be generated which can be fitted on the information obtained about the structural elements and the other spectral data of the unknown alkaloid. What can be predicted by means of biosynthetic reasoning is shown in Fig. 1. Terpenoid-indole alkaloids are all derived from one common precursor, strictosidine. After hydrolysis of this glucoside, an intermediate with several reactive groups is obtained. In fact, all combinations of aldehyde and amine functions are found within this class of alkaloids.

In Schemes 1 and 2, the general strategies for respectively identification of a known alkaloid and the structure determination of a new alkaloid are summarized.

3.1 Spectroscopic Methods

3.1.1 UV (Ultraviolet)

This is the oldest among the spectroscopic methods, and still an important tool in the identification of alkaloids, particularly for groups like indole and isoquinoline alkaloids which have quite a few different, characteristic chromophores, yielding information about the aromatic part of the molecule (Sangster and Stuart 1965). Some examples of spectra of different types of

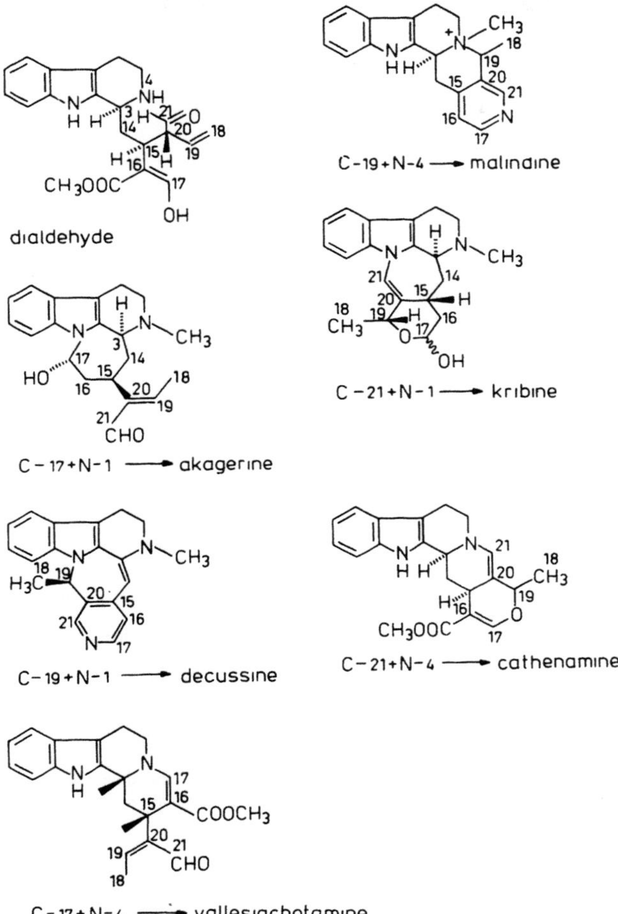

Fig. 1. Some alkaloids derived from strictosidine

indole alkaloids are given in Fig. 2. In Fig. 3, the effect of the position of substitution on the UV spectrum is illustrated. By measuring UV spectra at various pH, information can be obtained about the presence of phenolic groups.

3.1.2 IR (Infrared)

Originally quite an important tool in the structure elucidation of natural compounds, nowadays its use is limited. Because of its highly characteristic pattern of absorptions, IR is particularly useful for the confirmation of the

General strategy	**Example**

Isolation method
⇓
　　Alkaloid　　　　　　　　　　　　　　ca. 16.000 alkaloids are known

　⇓　　　　　　　　　　　　　　　　　　　　⇓
Chemotaxonomy (plant genus)　　　　　　*Strychnos*
　⇓　　　　　　　　　　　　　　　　　　　　⇓
　　Type of alkaloid　　　　　　　　　　ca. 350 alkaloids are known

　⇓　　　　　　　　　　　　　　　　　　　　⇓
TLC + colour reactions　　　　*HRf-values, colour with FeCl₃/HClO₄, CeSO₄/H₂SO₄*
UV　　　　　　　　　　　　　　*Chromophore*
MS　　　　　　　　　　　　　　*Molecular weight*
NMR　　　　　　　　　　　　　*Characteristic features*

　⇓　　　　　　　　　　　　　　　　　　　　⇓
Identification known alkaloid　　　　　Identification

Scheme 1. Strategy for the identification of alkaloids

Chemotaxonomy	→	type of alkaloid
UV	→	chromophore
MS	→	molecular weight, known fragments, simple derivative of known alkaloid (fragments +14, +16, +30, etc.)
¹HNMR	→	characteristic features, eventually complete assignment with the aid of 2D methods and nOe
¹³CNMR	→	characteristic features, functional groups, 2D-methods for e.g. C-H (long range) couplings
IR	→	functional groups

The structural elements found are, based on biosynthetic reasoning, combined into possible structures. These structures are compared with all spectral data. This will lead to a final proposal for a structure.

Scheme 2. Strategy for structure determination of novel alkaloids

identity of known compounds. In structure elucidation it can be useful in the identification of certain functional groups, e.g., carbonyl groups; but also for the determination of stereochemistry it can be a useful method, e.g., hydrogen bonding. In the case of heteroyohimbine alkaloids, the occurrence of Bohlman bands at about $2900\,\mathrm{cm}^{-1}$ is indicative of the stereochemistry of the C and D rings (Crab et al. 1971).

3.1.3 ORD and CD
(Optical Rotation Dispersion and Circular Dichromism)

Chiroptical methods have developed as major tools for solving the absolute stereochemistry in natural products (Scopes 1975). For example, in the case of indole alkaloids the configuration at C-2 and C-3 can be determined by

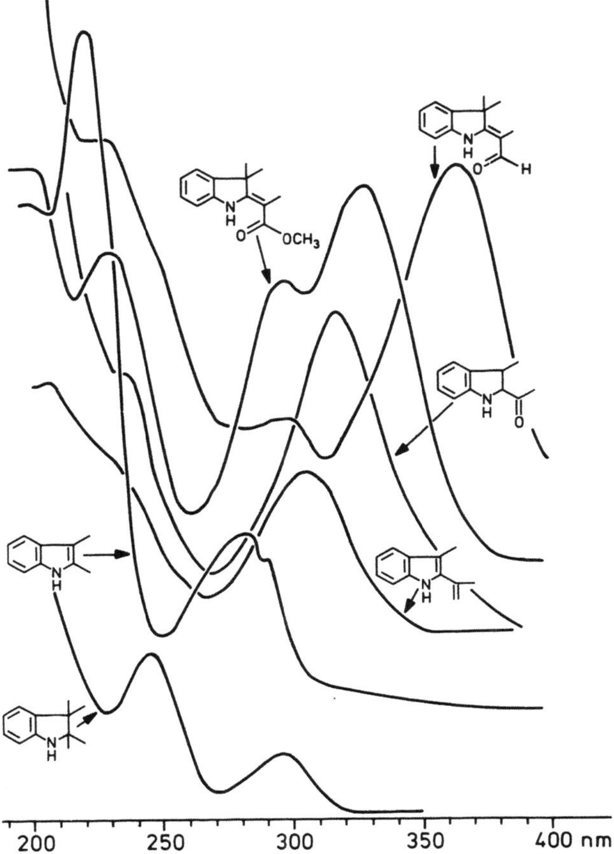

Fig. 2. UV spectra of some common indole chromophores (not normalized for concentrations)

these methods (e.g., Blaha et al. 1968, 1974a,b; Snow and Hooker 1978; Toth et al. 1980).

3.1.4 X-Ray Crystallography

The most powerful method, in the sense that it will give the complete structure and stereochemistry and conformation of a compound. The major constraint is that suitable crystals are needed. In the case of unknown compounds the other spectral data still have to be recorded to be able to identify the compound in future isolations. X-ray crystallography will thus only be applied after all other spectral data have been recorded, probably

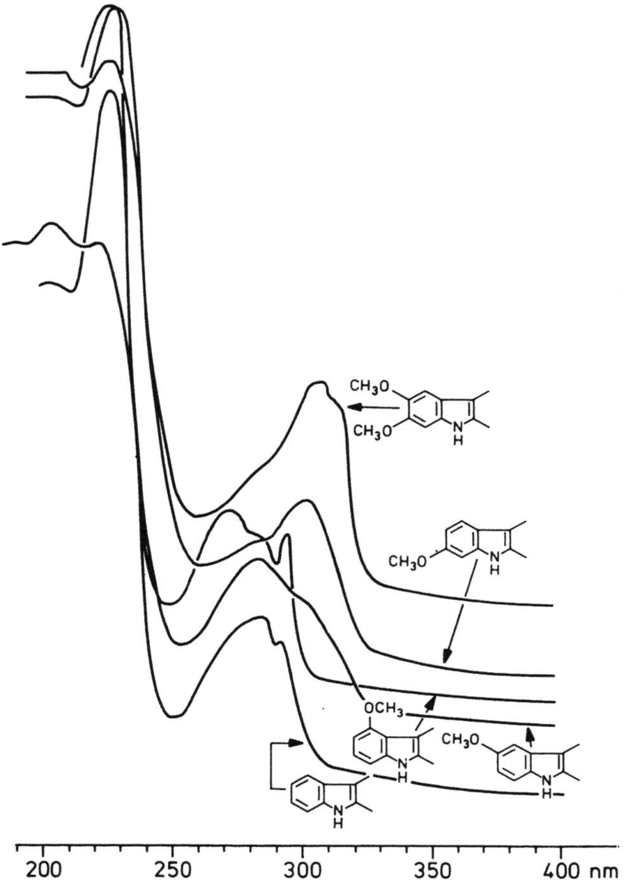

Fig. 3. UV spectra of some aromatic substituted indole alkaloids (not normalized for concentrations)

already resulting in the identification or structure determination of the unknown alkaloid.

3.1.5 MS (Mass Spectrometry)

In the past decade, a number of soft ionization modes have been developed which now make it possible in most cases to obtain the molecular weight. Even for compounds with very high molecular weights, like proteins, these can now be determined by means of MS. Ionization methods commonly used are fast atom bombardment (FAB), field desorption (FD) and chemical ionization (CI). Chemical ionization spectra can be recorded from both positive and negative ions, the combination of this information may sometimes be very helpful in the identification (see Chap. 3 on LC-MS). Even for labile alkaloids, molecular ions can be obtained by applying such soft ionization methods.

An important aspect of MS (especially electron impact) is fragmentation, which is characteristic for each compound. With methods like FAB and FD, this characteristic feature is often lost. Fragmentation patterns do give useful information about the structure of a compound. Loss of certain groups, or specific fragments for a certain class of compounds, are important information for the structure elucidation. Furthermore, with high resolution mass spectrometry, the elemental composition of the molecule can be obtained. The MS of alkaloids has been extensively reviewed by Hesse (1974) and Hesse and Bernhard (1975).

The direct coupling of MS with GC or HPLC turns these methods into very powerful tools in the identification of alkaloids at low levels in various biological materials (see separate chapters on GC-MS and LC-MS). GC-MS has the advantage that it is easier to obtain mass spectra also showing fragmentation, either using electron impact (EI) or chemical ionization (CI). In most of the available interfaces for LC-MS, no fragmentation spectra can be obtained; only the protonated molecule (M+1) can be observed in the positive ion mode.

3.1.6 NMR (Nuclear Magnetic Resonance)

Since its introduction in natural products chemistry in the early 1960s, NMR has developed as the most important tool in identification and structure elucidation. It was the first nondestructive method that gave direct information on the presence of certain functional groups such as methyl, amino, hydroxyl, methoxyl, double bonds, and aromatic protons. It also gave information about the relation between certain groups, through the couplings shown by the signals. As the first NMR spectrometer operated only at low magnetic field strength (typically 60 MHz), the resolution of the spectra was poor. For more complex molecules, the signals of the aliphatic protons could not be distinguished separately, and that part of the spectra was only useful as a fingerprint for identification purposes; but usually above 3 ppm more distinct signals could be observed, which were of great help in identifying certain structural elements.

The revision of the structure of tubocurarine in 1970, proving that, in fact, this alkaloid was a monoquaternary and not a bisquaternary alkaloid, clearly shows the role of NMR at that time (Everette et al. 1970).

In the 1970s, the introduction of Fourier transform (FT) NMR enabled for example ^{13}C-NMR, a further tool for structure elucidation. ^{13}C-NMR spectrometry was a major breakthrough, as it provided information on all the carbons of a molecule: the chemical shifts of the carbons gave information

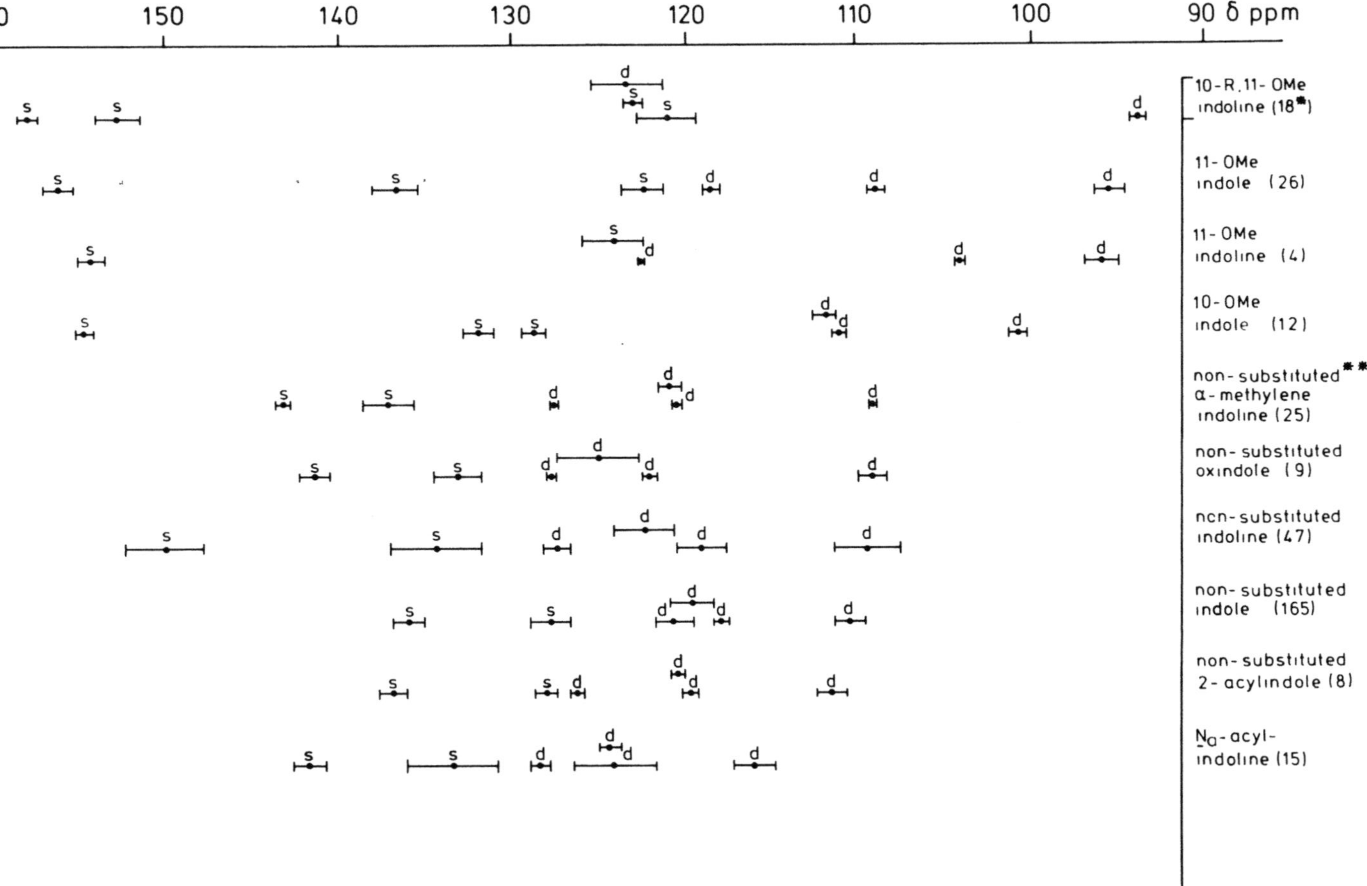

Scheme 3. Chemical shifts of the aromatic carbons in the most common chromophoric groups in indole alkaloids [d = doublet (CH), s = singlet (C)]

Table 1. Parameters which can be obtained by NMR spectroscopy

^{1}H chemical shifts
^{13}C chemical shifts

Coupling constants:
$^2J_{HH}$: geminal proton-proton coupling
$^3J_{HH}$: vicinal proton-proton coupling
long range proton-proton couplings
$^1J_{CH}$: direct carbon-proton coupling
$^2J_{CH}$: two-bond carbon-proton coupling
$^3J_{CH}$: three-bond carbon-proton coupling
longer range carbon-proton couplings

nOe: through space interactions between protons

about their chemical environment (e.g., substituents directly attached, or close in space) and the number of protons attached could also be determined.

Numerous papers have been published with ^{13}C-NMR data on alkaloids, as well as several reviews and a book with data on a number of alkaloids (Shamma and Hindenlang 1979). Based on these spectral data, it is possible to draw general conclusions about the chemical shifts of certain structural features, which might be helpful in structure elucidation. For example, the type of indole moiety of indole alkaloids can easily be determined on the basis of the shifts of the aromatic carbons, also the substituents and the substituent pattern can be deduced from these data (Verpoorte et al. 1984; Scheme 3).

The last decade high-resolution NMR, i.e., 300–600 MHz ^{1}H-NMR, became available as a routine method. Complete assignments are possible because of the improved resolution in combination with a variety of 2D-NMR methods enabling to record which protons couple and to determine nOe's between spatially close protons. In Table 1 the parameters which can be obtained from NMR spectroscopy are summarized. With the increased sensitivity also ^{15}N-NMR has become possible. Its use for structure elucidation is, however, limited.

3.1.7 ^{1}H-NMR

The NMR analysis of an unknown compound starts with a normal ^{1}H-NMR spectrum. Each proton in the molecule has a characteristic chemical shift and might display coupling constants with nearby protons. Information about functional groups, the amount of protons, and the position of protons relative to each other is obtained. Especially the vicinal proton-proton coupling constants are very useful for assignment of the stereochemistry of groups attached to cyclic systems, because the magnitude of this coupling depends on the dihedral angle between the protons according to a relation known as the Karplus equation.

The ^{1}H-NMR spectrum may give a lead by showing characteristic features already known from other alkaloids, e.g., the pattern of aromatic signals in indole alkaloids can be used to obtain information about the type of indole alkaloid involved (Schripsema and Verpoorte 1991).

A complicating factor in the interpretation of the ^{1}H-NMR spectra might be the fact that signals of different protons are overlapping. By using shift reagents, resolution of overlapping signals can be obtained. Trifluoroacetic acid was shown to be a useful shift reagent for alkaloids (Schripsema et al. 1986). In $CDCl_3$ solution it forms a soluble ion pair with the alkaloids. The shifts obtained are especially large for protons in the neighborhood of the protonated nitrogen, which gives further information about the structure and the stereochemistry of the alkaloid.

Another way to obtain resolution of overlapping signals might be the application of the HOHAHA technique, by which spectra from the separate spin systems can be obtained, or the measurement of 2D J-resolved spectra. These techniques will be discussed below.

3.1.8 ^{13}C-NMR

The next step in the structure elucidation is the measurement of a ^{13}C-NMR spectrum. Each carbon gives a signal at a shift characteristic for its molecular environment, e.g., aromatic, vinylic, with hetero-atomic substituents. Because the natural abundance of ^{13}C is only 1.1%, and the lower magnetogyric ratio, the sensitivity of ^{13}C-NMR is much lower than of ^{1}H-NMR.

To obtain a maximum sensitivity, the ^{13}C-NMR spectra are usually measured with proton decoupling. Distortionless enhancement by polarization transfer (DEPT) or attached proton test (APT) ^{13}C-NMR spectra enable the determination of the number of attached protons for each carbon. If sufficient material is available, also proton-coupled spectra can be measured, which contain all carbon-proton coupling information. The magnitude of $^3J_{CH}$, like that of $^3J_{HH}$, depends on the dihedral angle between the coupling atoms, and can therefore be very useful in determining stereochemical relations.

3.1.9 2D-NMR

A large number of two-dimensional NMR techniques are available nowadays. In the following, the most useful and most applied techniques will be discussed from the point of view of practical utility. References are given for the experimental details. An overview of the methods is given in Table 2.

Some general points about the representation of 2D spectra might be useful. In fact, these 2D spectra are three-dimensional, which means that when plotting them some information is lost. One way of plotting is the so-

Table 2. Overview of 2D-NMR techniques

Correlation:	$^1H-^1H$	$^1H-^{13}C$	$^{13}C-^{13}C$
	COSY	HETCOR	INADEQUATE
	HOHAHA	HMQC	
	NOESY	HMBC	
		COLOC	

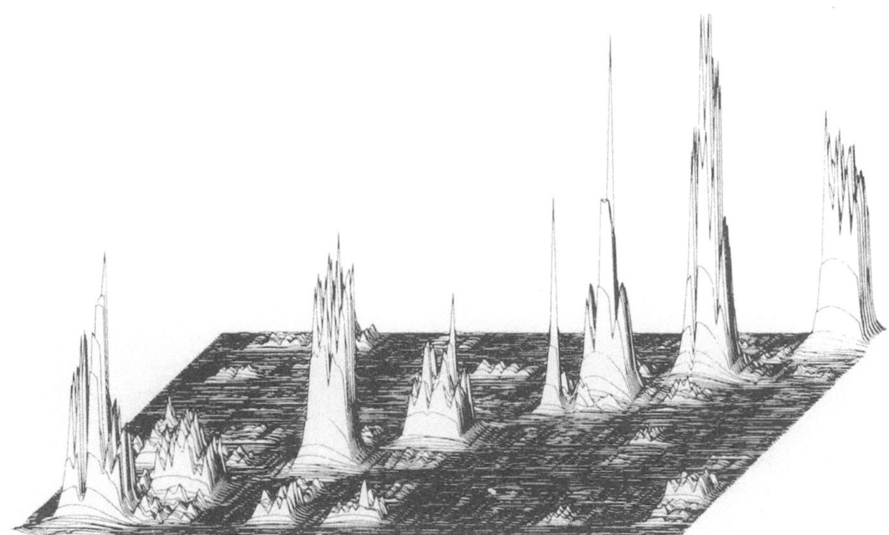

Fig. 4. Stacked plot from a NOESY spectrum of tubotaiwine (Fig. 7) (CDCl$_3$, 300 MHz)

called stacked plot (Fig. 4). But often peaks are not visible because they are hidden behind other peaks. The most common way of plotting is the so-called contour plot (e.g., Fig. 5). Here the contours of the signals are drawn at selected levels. In this type of plot one has to be very careful with the selection of the levels. Taking it too low means a lot of noise in the spectrum, while taking it too high means that valuable signals might be lost. Therefore, care should be taken in the interpretation of these contour plots, signals may be present below the lowest level of the plot and seemingly interesting signals might be due to noise.

COSY (Correlated Spectroscopy). This type of spectrum shows which protons couple with which protons (Aue et al. 1976). On each axis there is the ^{1}H-NMR spectrum and the cross-peaks indicate a coupling between the protons (Fig. 5). This spectrum is especially useful when a number of identical coupling constants are present in several signals. Care should be

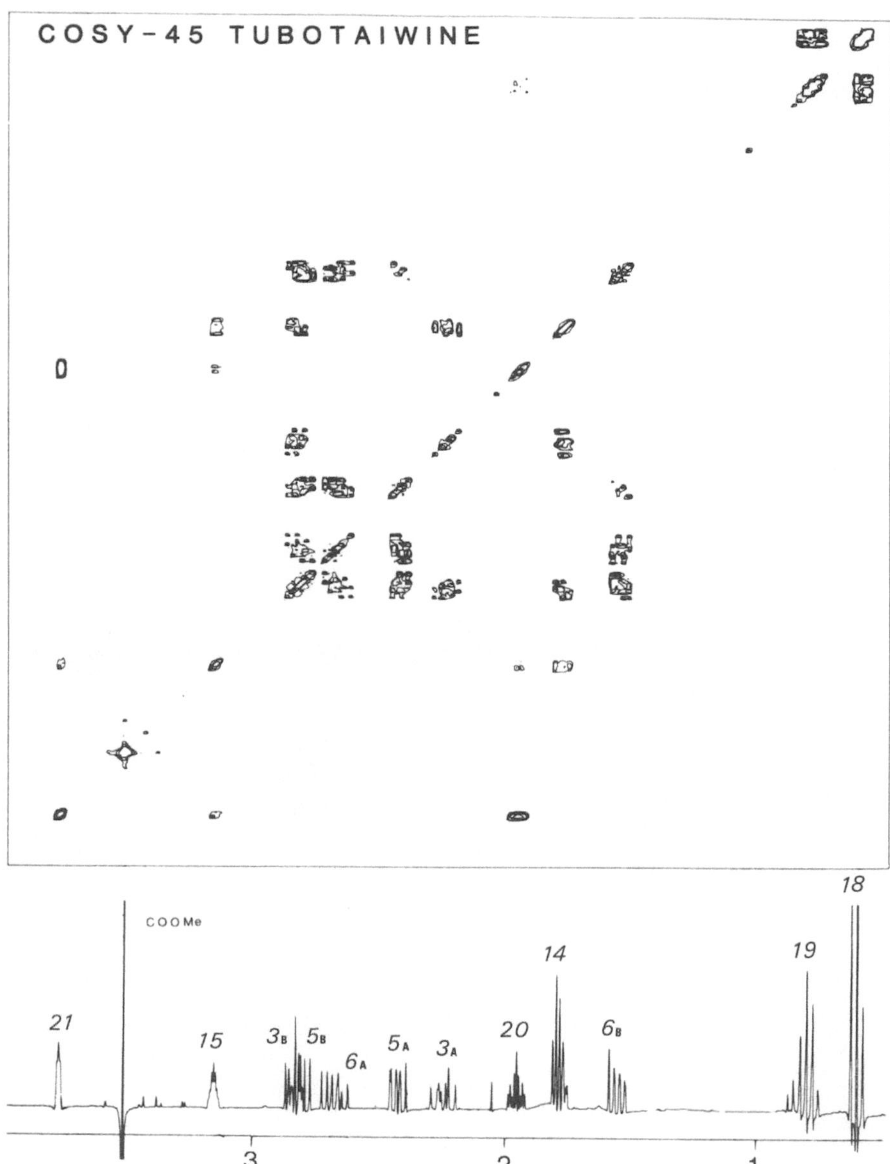

Fig. 5. COSY spectrum of tubotaiwine (Fig. 7) (CDCl₃, 300 MHz)

taken in the interpretation of cross-peaks when signals are overlapping. The experiment can also be optimized for long-range couplings.

HOHAHA (Homonuclear Hartmann-Hahn Spectroscopy). This method is based on spin propagation (Braunschweiler and Ernst 1983; Summers et al.

1986). During a propagation delay in the pulse sequence, the magnetization propagates through a proton-coupling network. If a long delay is taken (about 100–200 ms), the complete spin system will be visible. By choosing shorter delays, assignment of the signals is possible, e.g., with a delay of 40 ms only the directly coupled protons will become visible. The experiment can be performed two-dimensionally, but can also be done one-dimensionally, giving selective subspectra (Davis and Bax 1985).

NOESY (Nuclear Overhauser Enhancement Spectroscopy). This technique depends on the occurrence of dipolar cross-relaxation (Bodenhausen et al. 1984). This so-called nOe effect depends on the distance through space and is independent of direct bonding. In this way it is a powerful technique to establish the stereochemistry and conformation of molecules. In Fig. 6 the

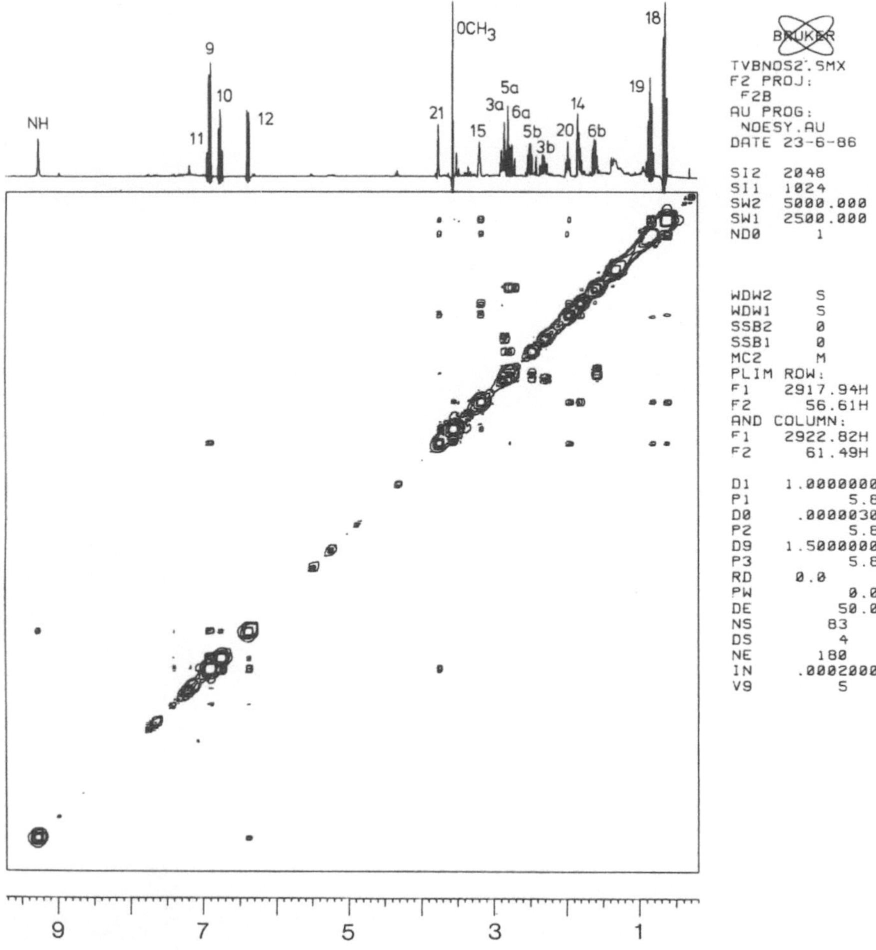

Fig. 6. NOESY spectrum of tubotaiwine (Fig. 7) (C_6D_6, 300 MHz)

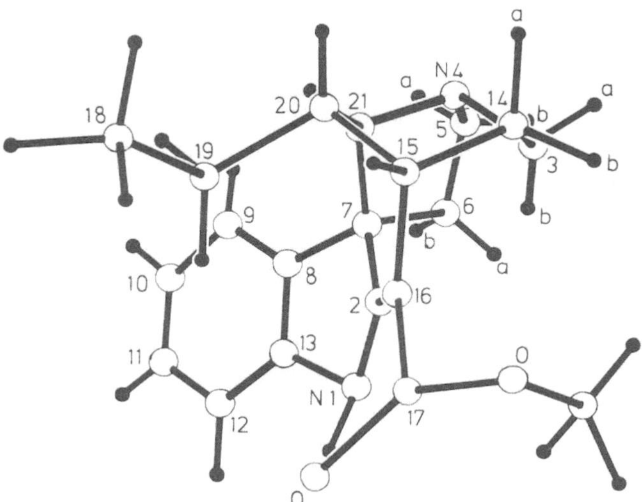

Fig. 7. Stereochemistry of tubotaiwine

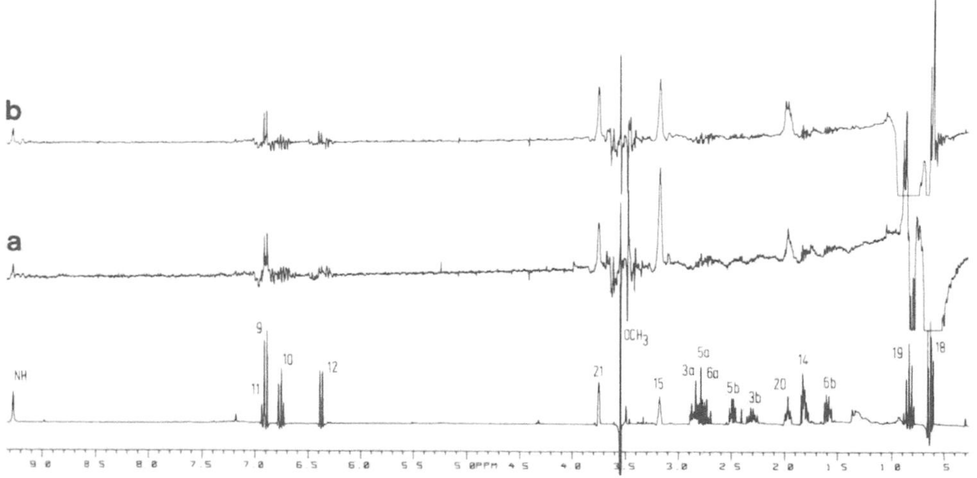

Fig. 8a,b. NOe difference spectra of tubotaiwine (Fig. 7). **a** Irradiated at H-19. **b** Irradiated at H-18 (C_6D_6, 300 MHz)

NOESY spectrum of the indole alkaloid tubotaiwine (Fig. 7) is shown. It shows some very useful cross-peaks, e.g., between H-21 and H-9, the indole NH and H-12, and between H-14 and H-20. The one-dimensional nOe difference spectrum (Neuhaus 1983) is sometimes preferable because of its higher sensitivity (Fig. 8). However in the case of signals with very close chemical shifts the 2D experiment is clearly better, e.g., the nOe effect between H-20 and H-14a in Fig. 6.

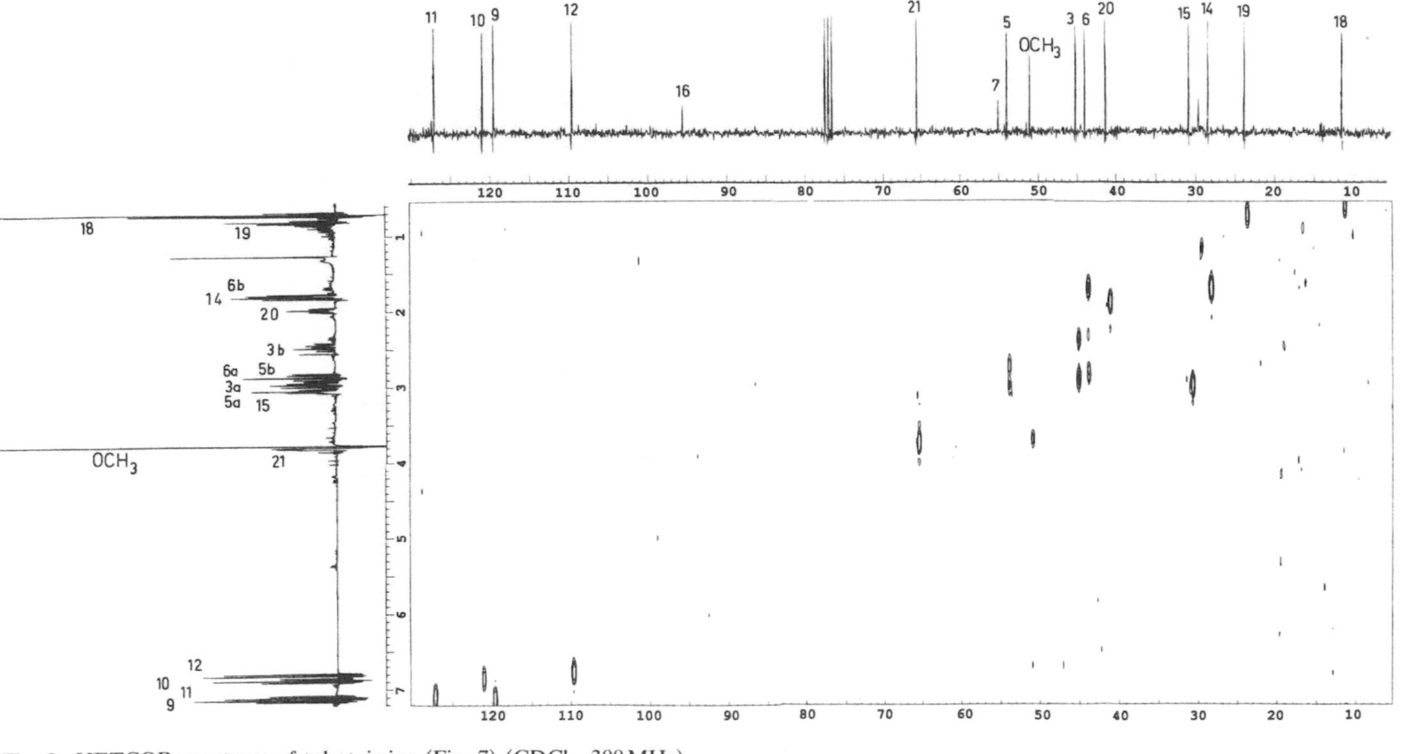

Fig. 9. HETCOR spectrum of tubotaiwine (Fig. 7) (CDCl$_3$, 300 MHz)

HETCOR (Heteronuclear Correlation). A 2D-heteroCOSY spectrum shows for all protons to which carbon they are attached (Bax 1983). On one axis is the ^{1}H-NMR spectrum while on the other one is the ^{13}C-NMR spectrum. Signals indicate a direct coupling of the proton with the carbon (Fig. 9). Geminal protons can easily be recognized.

HMQC (Heteronuclear Multiple-Quantum Coherence). This type of spectrum (Summers et al. 1986) gives essentially the same information as the HETCOR spectrum, but with higher sensitivity (about 15 times more). However, a special probe is needed to obtain these spectra.

HMBC (Heteronuclear Multiple Bond Connectivity). This spectrum (Bax and Summers 1986; Summers et al. 1986) has, like the HETCOR and HMQC, on one axis the ^{1}H-NMR spectrum and on the other the ^{13}C-NMR spectrum. Cross-peaks are shown due to $^2J_{CH}$ and $^3J_{CH}$ and sometimes $^4J_{CH}$ long-range couplings. It is a very useful method to connect different spin systems. The advantage of this method above the long-range HETCOR is the much higher sensitivity due to the detection through protons instead of carbons.

COLOC (Correlation Spectroscopy via Long-Range Couplings). Long-range proton-carbon couplings can be selectively observed in this technique (Kessler et al. 1984, 1985). It gives information similar to that of the HMBC spectrum, but with lower sensitivity. As with HMBC and long-range HETCOR, it can be a useful technique to connect different spin systems.

INADEQUATE (Incredible Natural Abundance Double Quantum Transfer Experiment). On one axis one has the ^{13}C-NMR spectrum while on the other axis the double quantum frequencies are present (Englert 1985). A direct carbon-carbon coupling is indicated by a pair of doublets at a certain double quantum frequency. In this experiment, in fact the ^{13}C satellites of ^{13}C signals are observed. It must be clear that this method has a low sensitivity, because two adjacent ^{13}C nuclei are required.

J-Resolved Spectroscopy. In J-resolved spectra, the chemical shifts are displayed on one axis while on the other axis the coupling information is displayed. These spectra can be obtained both for protons and carbon-13. They can be useful in the case of crowded spectra.

4 Strategy

During the isolation of an alkaloid, usually data are obtained on TLC behavior and UV. The first step then is to record a ^{1}H-NMR spectrum. As discussed above, this is generally sufficient to identify known alkaloids. In

Table 3. Estimations of the amount of compound needed to obtain a certain type of spectrum in an overnight experiment. The quantities needed are further dependent on a number of factors, e.g. the field strength of the apparatus to be used, the tuning, and the molecular weight of the compound

Technique	Amount (mg)
Normal 1D ^{1}H-NMR	0.005
Decoupled ^{13}C-NMR	1
COSY	0.1
HOHAHA	0.5
NOESY	5
HETCOR	5
HMQC	1
HMBC	5
COLOC	30
INADEQUATE	100
J-resolved ^{1}H	0.5
J-resolved ^{13}C	10

the case of unknowns, more spectra need to be obtained. Generally, one of the main problems is the amount of compound available, and the choice of methods depends largely on this. Another constraint is the type of apparatus available. With higher fields, a higher sensitivity is obtained and for certain techniques special probes are needed, e.g., for HMQC and HMBC an inverse probe. Table 3 gives the above-mentioned NMR methods, with an estimation of the amount of compound needed to obtain a good spectrum in an overnight experiment.

The following procedure will usually be sufficient to solve any structure:

1. Normal ^{1}H-NMR spectrum
2. Decoupled ^{13}C-NMR spectrum
3. COSY spectrum
4. HETCOR or HMQC spectrum
5. COLOC or HMBC spectrum
6. NOESY spectrum or 1D nOe difference spectra.

5 Future Prospects

In recent years, computer programs have been developed which are capable of proposing possible structures on the basis of NMR data, i.e., chemical shifts, H–H couplings, C–C couplings, C–H couplings, long range couplings, and nOe's (Nuzillard and Massiot 1991). Such systems have already been tested successfully on compounds with complex structures like strychnine and azadirachtin. In the near future, such methods will certainly revolutionize structure elucidation in the sense that it will be almost completely computerized, the scientist retaining only the task of choosing between possible structures by perhaps simply using chemotaxonomic and biosynthetic reasoning.

It may be that in the more distant future, the exact calculation of chemical shifts will become possible, thus enabling a complete structure analysis by a single ^{1}H-NMR spectrum.

References

Anderson LA, Doggett NS, Ross MSF (1977) The non-specificity of Dragendorff's reagent in thin layer chromatography. Planta Med 32:125–129

Aue WP, Bartholdi E, Ernst RR (1976) Two-dimensional spectroscopy. Application to nuclear magnetic resonance. J Chem Phys 64:2229–2246

Baerheim Svendsen A, Verpoorte R (1983) Chromatography of alkaloids. Part A. Thin-layer chromatography. J Chromatogr Libr vol 23A. Elsevier, Amsterdam

Bax A (1983) Broadband homonuclear decoupling in heteronuclear shift correlation NMR spectroscopy. J Magn Reson 53:517–520

Bax A, Summers MF (1986) ^{1}H and ^{13}C assignments from sensitivity-enhanced detection of heteronuclear multiple-bond connectivity by 2D multiple quantum NMR. J Am Chem Soc 108:2093–2094

Beke D (1963) Heterocyclic pseudo bases. In: Karitzky AR (ed) Advances in heterocyclic chemistry I. Academic Press, New York, p 167

Bisset NG, Casinovi CG, Galeffi C, Marini-Bettolo GB (1965) Su alcune 16-alcossi-stricnine. Ric Sci 35 (II-B):273–274

Blaha K, Kavkova K, Koblicova Z, Trojanek J (1968) On alkaloids XX. Absolute configuration of alkaloids of eburnane series. Optical rotation dispersion. Coll Czech Chem Commun 33:3833–3840

Blaha K, Koblicova Z, Trojanek J (1974a) Chiroptical investigations on the Iboga and Voacanga alkaloids. Coll Czech Chem Commun 39:2258–2266

Blaha K, Koblicova Z, Trojanek J (1974b) Chiroptical properties of the 5,16-cyclocorynane type alkaloids. Coll Czech Chem Commun 39:3168–3176

Bodenhausen G, Kogler H, Ernst RR (1984) Selection of coherence-transfer pathways in NMR pulse experiments. J Magn Reson 58:370–388

Braunschweiler L, Ernst RR (1983) Coherence transfer by isotropic mixing: application to proton correlation spectroscopy. J Magn Reson 53:521–528

Brossi A (ed) (1983–1992) The alkaloids, vols 21–40. Academic Press, New York

Conway WD (1990) Counter current chromatography. Apparatus, theory and applications. VCH, New York

Crab TA, Newton RF, Jackson D (1971) Stereochemical studies of N-bridgehead compounds by spectral means. Chem Rev 71:109–126

Davis DG, Bax A (1985) Simplification of ^{1}H NMR spectra by selective excitation of experimental subspectra. J Am Chem Soc 107:7197–7198

Englert G (1985) NMR of carotenoids – new experimental techniques. Pure Appl Chem 57:801–820

Everette AJ, Lowe LA, Wilkinson S (1970) Revision of the structures of (+)-tubocurarine chloride and (+)-chondrocurine. J Chem Soc Chem Commun: 1020–1021

Farnsworth NR, Blomster RN, Damratoski D, Meer W, Cammarato LV (1964) Studies on Catharanthus alkaloids. VI evaluation by means of thin-layer chromatography and ceric ammonium sulfate spray reagent. Lloydia 27:302–314

Formacek V, Kubeczka K-H (1982) Essential oils analysis by capillary gas chromatography and carbon-13 NMR spectroscopy. John Wiley, Chichester

Glasby JS (1975) Encyclopedia of alkaloids, vols 1 and 2. Plenum, New York

Hermans-Lokkerbol A, Verpoorte R (1986) DCCC of alkaloids. The influence of pH-gradients and ion-pair formation on the retention of alkaloids. Planta Med 52:299–302

Hesse M (1974) Progress in mass spectrometry. Vol 1, parts 1 and 2. Mass spectrometry of indole alkaloids. Verlag Chemie, Weinheim

Hesse M, Bernhard HO (1975) Progress in mass spectrometry. Vol 3. Mass spectrometry of alkaloids. Verlag Chemie, Weinheim

Jordan W, Scheuer PJ (1965) Rapid separation of alkaloids from plant material by anion exchange of Mayer's complex. J Chromatogr 19:175–176

Kessler H, Griesinger C, Zarbock J, Loosli HR (1984) Assignment of carbonyl carbons and sequence analysis in peptides by heteronuclear shift correlation via small coupling constants with broadband decoupling in t_1 (COLOC). J Magn Reson 57:331–336

Kessler H, Bermel W, Griesinger C (1985) Recognition of NMR proton spin systems of cyclosporin A via heteronuclear proton-carbon long-range couplings. J Am Chem Soc 107:1083–1084

Manske RHF (ed) (1955–1977) The alkaloids, vols 6–16. Academic Press, New York

Manske RHF, Holmes HL (eds) (1950–1955) The alkaloids, vols 1–5. Academic Press, New York

Manske RHF, Rodrigo R (eds) (1979–1981) The alkaloids, vols 17–20. Academic Press, New York

Neuhaus D (1983) A method for the suppression of selective population transfer effects in NOE difference spectra. J Magn Reson 53:109–114

Nuzillard J-M, Massiot G (1991) Computer-aided spectral assignment in nuclear magnetic resonance spectroscopy. Anal Chim Acta 242:37–41

Pelletier SW (ed) (since 1983) Alkaloids: chemical and biological perspectives, vols 1–6. John Wiley, New York

Phillipson JDP, Bisset NG (1972) Quaternisation and oxidation of strychnine and brucine during plant extractions. Phytochemistry 11:2547–2553

Popl M, Fähnrich J, Tatar V (1990) Chromatographic analysis of alkaloids. Marcel Dekker, New York

Sangster AW, Stuart KL (1965) Ultra-violet spectra of alkaloids. Chem Rev 65:69–130

Schripsema J, Verpoorte R (1991) Rapid identification of trace amounts of indole alkaloids: analysis of the aromatic pattern from the ^{1}H-NMR spectrum. In: Atta-ur-Rahman (ed) Studies in natural products chemistry, vol 9. Elsevier, Amsterdam, pp 163–199

Schripsema J, Verpoorte R, Baerheim Svendsen A (1986) Trifluoroacetic acid, a ^{1}H-NMR shift reagent for alkaloids. Tetrahedron Lett 27:2523–2526

Scopes PM (1975) Applications of the chiroptical methods to the study of natural products. Fortschr Chem Org Naturst 32:167–265

Shamma M, Hindenlang DM (1979) Carbon-13 NMR shift assignments of amines and alkaloids. Plenum, New York

Siek TJ, Eichmeier LS, Caplis ME, Esposito FE (1977) The reaction of normeperidine with an impurity in chloroform. J Anal Toxicol 1:211–214

Snow JW, Hooker TM (1978) The chiroptical properties of strychnine alkaloids: strychnine, β-colubrine, brucine and their dihydro derivatives. Can J Chem 56:1222–1230

Southon IW, Buckingham J (1989) Dictionary of alkaloids. Chapman and Hall, London

Summers MF, Marzilli LG, Bax A (1986) Complete ^{1}H and ^{13}C assignments of coenzyme B_{12} through the use of new two-dimensional NMR experiments. J Am Chem Soc 108:4285–4294

Toth G, Clauder O, Gesztes K, Yemul SS, Snatzke G (1980) Circular dichroism of indole alkaloids. Part I. Vincane alkaloids. J Chem Soc Perkin Trans 2:701–703

Van Beek TA, Verpoorte R, Baerheim Svendsen A (1984) Identification of *Tabernaemontana* alkaloids by means of thin-layer chromatography and chromogenic reactions. J Chromatogr 298:289–307

Van der Heijden R, Hermans-Lokkerbol A, Verpoorte R, Baerheim Svendsen A (1987) Pharmacognostical studies on *Tabernaemontana* XX. Ion-pair droplet counter-current chromatography of indole alkaloids from suspension cultures. J Chromatogr 396: 410–415

Verpoorte R, Baerheim Svendsen A (1976) Extraction of alkaloids from the bark of *Strychnos dolichothyrsa*. Pharm Weekbl 111:833–836

Verpoorte R, Baerheim Svendsen A (1984) Chromatography of alkaloids. Part B: Gas-liquid chromatography and high-performance liquid chromatography. J Chromatogr Libr vol 23A. Elsevier, Amsterdam

Verpoorte R, Visser MGLM, Baerheim Svendsen A (1983) False positive alkaloid reactions with iodoplatinate reagent compared with Dragendorff's reagent. Fitoterapia 54:127–131

Verpoorte R, Van Beek TA, Riegman RLM, Hylands PJ, Bisset NG (1984) Aromatic chemical shifts in ar-hydroxy- and -methoxysubstituted indole alkaloids; reference data and substituent-induced chemical shifts for ten different chromophoric groups. Org Magn Reson 22:328–335

Inverse-Detected 2D-NMR Applications in Alkaloid Chemistry

G.E. MARTIN and R.C. CROUCH

1 Introduction

Clearly the development and dissemination of two-dimensional NMR techniques has had a profound impact in natural products structure elucidation. Some techniques, COSY and variants of the ^{13}C-detected heteronuclear chemical shift correlation (variously referred to as HETCOR, HC-COSY, etc.) experiment, have been widely used by the natural products chemistry community. Inverse-detected heteronuclear shift correlation techniques are becoming recognized as a powerful adjunct to the COSY experiment and a replacement for their less sensitive and, in some cases, less versatile ^{13}C-detected predecessor experiments (Martin and Crouch 1991).

Benefits of increased sensitivity with inverse-detection are certainly appreciated by the natural products chemistry community. There seem, however, to be misperceptions associated with inverse-detection which may have impeded application in some laboratories. As the categoric name implies, inverse-detected experiments detect proton rather than carbon. Older spectrometers, unless they have been specifically modified to allow the application of ^{13}C pulses from the decoupler, may not be able to perform inverse-detected experiments (Zektzer et al. 1988). Older instruments capable of delivering high power ^{13}C decoupler pulses may still be limited in the range of inverse-detected experiments that they can perform if they cannot be programmed to perform some form of broadband heteronucleus, e.g., WALTZ, decoupling. Beyond instrumental limitations, there also seems to be a misconception on the part of some natural products chemists that an inverse-detection probe must be available to perform the experiments. Not true! Inverse-detected heteronuclear shift correlation experiments can be performed with any probe capable of performing proton-carbon heteronuclear shift correlation experiments, albeit with somewhat lower sensitivity than would be available if an inverse-geometry probe were employed. Despite lower intrinsic sensitivity arising from the use of a normal geometry probe, there is still a significant sensitivity advantage to be gained from performing the proton- rather than carbon-detected experiment. Indeed, even if the instrument being used can only perform the long-range HMBC experiment of Bax and Summers (1986), there is still an immense sensitivity advantage that should be exploited.

Given this introduction, we will digress to briefly discuss the historical development of inverse-detected experiments – which are much older than

Modern Methods of Plant Analysis, Volume 15
Alkaloids (ed. by Linskens/Jackson)
© Springer-Verlag Berlin Heidelberg 1994

many natural products chemists might realize. A description of the one-bond correlation experiment follows, and the function of the various constituents of the pulse sequences are discussed. Parameters and considerations that the authors have found useful in performing these experiments are included. Some of the numerous alternatives to the HMQC experiment for one-bond (direct) heteronuclear shift correlation are described. A more comprehensive treatment of the methods available is presented in the review of Martin and Crouch (1991). Inverse-detected heteronuclear multiple bond correlation (HMBC) experiments and heteronuclear correlation experiments with homonuclear relay and/or isotropic mixing periods are presented next. New directions, e.g., the development and utilization of one-dimensional and F_1 region-selected analogs of inverse-detected two-dimensional NMR experiments are the last experimental techniques to be considered. Next, sample requirements, which have changed significantly from when this chapter was first written to when it was edited, are discussed. Finally, applications of inverse-detected two-dimensional NMR experiments in the elucidation of alkaloid structures are surveyed. As spectroscopists, we have arbitrarily subdivided terrestrial and marine alkaloids with further sub-grouping of potential assignment/structural problems for which spectroscopic solutions now exist, but which have not yet arisen in the field of alkaloid chemistry.

2 Historical Background

Some readers may be surprised to learn that inverse-detected two-dimensional NMR experiments are not a recent addition to the pulse sequence libraries of the NMR spectroscopist. The HMQC (Heteronuclear Multiple-Quantum Correlation) experiment of Bax and Subramanian (1986), now in widespread use, was preceded by the pioneering work of Müller (1979) 7 years earlier. Müller described a pulse sequence that is not substantially different from the HMQC experiment of Bax and Subramanian (1986).

Following Müller's seminal work, a number of other research groups reported early efforts to utilize inverse-detected heteronuclear correlation experiments (Bodenhausen and Ruben 1980; Bax et al. 1983a,b; Bendall et al. 1983; Live et al. 1984, 1985; Frey et al. 1985; Otvos et al. 1985; Bax and Subramanian 1986; Brühwiler and Wagner 1986; Mueller et al. 1986; Wilde et al. 1986; Leupin et al. 1987; Sklenar and Bax 1987). It is beyond the scope of this chapter to consider all of the interesting work reported by these authors. We will, however, consider several experiments now in use which are applicable to alkaloid structure problems (see Sect. 4).

3 Suppression of Unwanted Magnetization Arising from $^1H-^{12}C$ Species

A problem inherent to one-bond, inverse-detected experiments was the suppression of the unwanted 99% of the signal arising from protons directly bound to ^{12}C and thus of no interest spectroscopically. A number of approaches to the problem of suppressing the $^1H-^{12}C$ signal are plausible and many were tried. In principle, it is possible to saturate the protons and then transfer the signal from the enhanced low gamma nuclide back to the proton to which it is directly bound for detection (Neuhaus et al. 1985). Unfortunately, the sensitivity advantage gained by detecting protons rather than the heteronucleus is partially lost using this approach.

The approach to the suppression of the unwanted $^1H-^{12}C$ signal described by Bax and Subramanian (1986; see Fig. 1) has probably seen the most widespread utilization. Utilizing the discriminatory capability of the BIRD pulse (Garbow et al. 1982), the component of proton magnetization arising from $^1H-^{12}C$ species can be selectively inverted. By simply allowing time for longitudinal relaxation, typically 0.3–0.5 s for small organic molecules, the signal for $^1H-^{12}C$ species is effectively nulled prior to the creation of heteronuclear multiple-quantum coherence. Practically, this technique works quite well and requires no prior knowledge of the T_1 relaxation time of the protons involved in most instances where one-bond heteronuclear correlation experiments are being performed.

Although a BIRD pulse works quite well to suppress unwanted $^1H-^{12}C$ magnetization when small molecules are being studied, BIRD pulses cannot be used with larger molecules, because the negative nOe would attenuate signals arising from protons attached to ^{13}C. The pulse sequences described by Sklenar and Bax (1987), one by Zuiderweg (1990), and others, circumvent these problems and can be effectively utilized with macromolecules. However, because virtually all alkaloids may be categorized as "small" molecules, pulse sequences such as HMQC (Bax and Subramanian 1986) and others discussed below may be employed without considering problems inherent to large molecules.

4 Inverse-Detected One-Bond Heteronuclear Shift Correlation Experiments

At present, several experiments are available for inverse-detected one-bond heteronuclear shift correlation. The HMQC experiment described by Bax and Subramanian (1986) has probably been most widely employed. Alternatives, however, are available in the form of DEPT-HMQC (Kessler et al. 1989b) and the HSQC or so-called Overbodenhausen experiment (Bodenhausen and Ruben 1980). For alkaloids with highly congested proton spectra, DEPT-HMQC may be a useful alternative to HMQC, because it allows the acquisition of edited correlation spectra. For investigators interested in correlation of protons to alkaloidal nitrogen atoms via one or two bonds, HSQC or a doubly refocused variant may be the preferred choice.

4.1 HMQC

Quite probably, HMQC (Bax and Subramanian 1986) is the most widely employed inverse-detected heteronuclear chemical shift correlation experi-

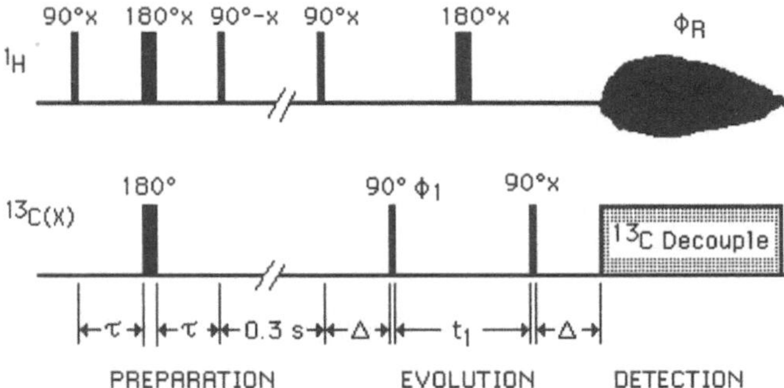

Fig. 1. HMQC pulse sequence of Bax and Subramanian (1986). The BIRD pulse in the preparation period selectively inverts the ^{1}H–^{12}C component of proton magnetization, which is nulled by the following delay of 0.3–0.5 s. Heteronuclear multiple-quantum coherence is created by the 90° ^{1}H pulse followed $1/2(^1J_{CH})$ (Δ) later by the 90° ^{13}C pulse. Zero- and double-quantum coherence terms evolve during the first half of the t_1 period and are interchanged by the 180° proton pulse midway through the period. The 180° ^{1}H pulse also serves to remove proton chemical shift evolution. The final 90° ^{13}C pulse converts the evolved multiple-quantum coherence to observable proton single-quantum coherence that, after refocusing by the Δ delay, may be heteronucleus broadband decoupled and detected

ment in use today. The HMQC pulse sequence is shown in Fig. 1. The BIRD pulse initiating the experiment is used to invert magnetization from ^{1}H–^{12}C species followed by a null delay to allow for longitudinal relaxation of the inverted magnetization component. Heteronuclear zero- and double-quantum coherences are created by the 90° ^{1}H and ^{13}C pulses following the null delay. The duration of the delay, Δ, separating the proton and carbon pulses is optimized as a function of $1/2(^1J_{CH})$. In cases where a compromise between aliphatic and aromatic carbons is necessary, we generally find it convenient to assume an average of 140 Hz for the $^1J_{CH}$ coupling constant. Heteronuclear zero- and double-quantum coherences evolve during the first half of the evolution period, $t_{1/2}$. The 180° proton pulse applied midway through the evolution period interchanges zero- and double-quantum terms in the density matrix and also serves to refocus ^{1}H chemical shift evolution so that the detected proton signal will be labeled only with the carbon chemical shift frequency during t_1. Evolved heteronuclear multiple-quantum coherence is converted to observable proton single-quantum coherence by the final 90° pulse in the sequence. When recreated, proton single-quantum coherence is antiphase and must be refocused during a second fixed delay, Δ, after which broadband heteronuclear decoupling and detection may be initiated.

4.2 DEPT-HMQC

DEPT-HMQC (Kessler et al. 1989b) combines the editing capabilities of the DEPT experiment (Doddrell et al. 1982) with the high sensitivity inherent to inverse-detected heteronuclear correlation experiments. When dealing with the often complex and congested proton NMR spectra of alkaloids, the editing capability of DEPT-HMQC may be advantageous. As the DEPT-HMQC experiment is discussed in more detail in the review of Martin and Crouch (1991), we will not consider it in depth here.

4.3 HSQC

The HSQC or "Overbodenhausen" experiment (Bodenhausen and Ruben 1980) shown in Fig. 2 differs from HMQC and DEPT-HMQC experiments because HSQC is single-quantum-based. Recently, HSQC has received considerable attention from peptide and protein NMR spectroscopists as a base for the development of yet more sophisticated two- and three-dimensional NMR experiments (see Kay et al. 1990 for examples). Probably the greatest advantage inherent to the use of single-quantum-based techniques accrues from the retained freedom to apply homonuclear proton decoupling during periods of heteronuclear evolution. Homonuclear decoupling applied during the evolution period of the HSQC experiment reduces heteronuclear linewidth, thereby affording improved resolution, which can be quite significant with larger molecules.

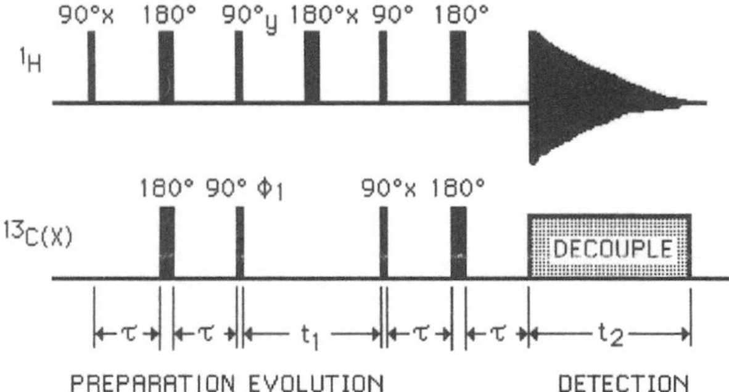

Fig. 2. The HSQC or "Overbodenhausen" experiment of Bodenhausen and Ruben (1980) is shown. An INEPT-type magnetization transfer to the heteronucleus occurs with the first 90° ^{13}C(X) pulse; the heteronuclear component of magnetization is then allowed to evolve. Proton chemical shift evolution can be removed by either applying a 180° pulse midway through the evolution period or by decoupling during evolution. Magnetization is then transferred back to proton, refocused, and detected with heteronucleus decoupling as shown

The applicability of the HSQC experiment in natural product structure elucidation studies remains to be evaluated. To the best of our knowledge, no natural product structure elucidation studies have been reported using HSQC rather than HMQC to establish direct proton-carbon chemical shift correlations. This will undoubtedly change in time. A modification of the HSQC experiment, inverted direct response HSQC-TOCSY, has, however, been reported using strychnine as a model compound (Domke 1991). In the case of congested proton spectra, there may be advantages inherent to the use of HSQC instead of HMQC.

4.4 HMQC Using Spin-Locking Fields

Another one-bond heteronuclear inverse-detected correlation technique worth mention here is that of Spitzer et al. (1989). Spitzer's experiment, for which there is no acronym, utilized proton spin-locking fields during the evolution period and prior to detection to suppress proton precession. Consequently, magnetization components are modulated solely by the ^{13}C precession frequency of interest, which should correspondingly provide better F_1 resolution. Homonuclear and heteronuclear scalar couplings are also eliminated, thereby obviating the need for refocusing pulses and the BIRD pulse used at the beginning of the HMQC experiment. Spitzer's work is also noteworthy because the experiment was demonstrated using the alkaloid deoxyvinblastine. No other applications of Spitzer's experiment have been reported.

4.5 Comparison of the Available Inverse-Detected One-Bond Heteronuclear Correlation Experiments

No direct comparison of HMQC and HSQC is available in the literature for any small molecule. Using the complex alkaloid cryptospirolepine (1), whose proton and carbon NMR spectra are reasonably congested (Tackie et al. 1993), HMQC and HSQC spectra were acquired, processed, and plotted

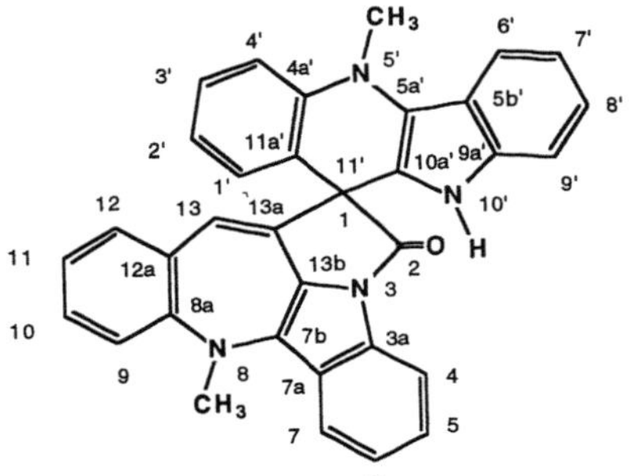

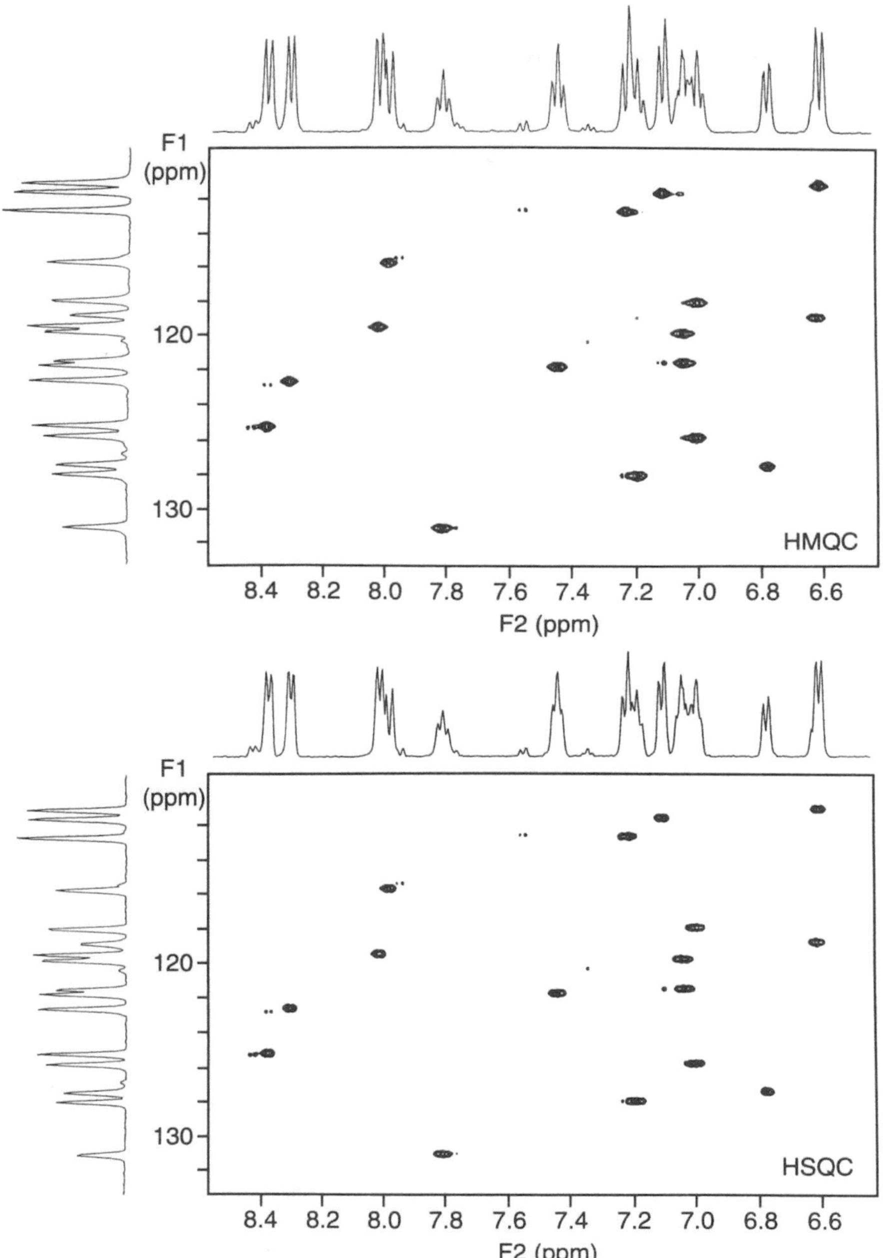

Fig. 3. Comparison of 500 MHz spectra of the aromatic region of the alkaloid cryptospirolepine (**1**). Both spectra were recorded under identical conditions using 2048 × 192 data points. A total of 32 transients were accumulated per file. The data were processed identically using Gaussian and cosine weighting in the F_2 and F_1 frequency domains, respectively. The HSQC (Bodenhausen and Ruben 1980) spectrum is shown in the *top panel*, the corresponding HMQC (Bax and Subramanian 1986) spectrum is shown in the *bottom panel*. Both spectra are flanked by projections to allow an evaluation of the comparative resolution in the two spectra

identically, and are shown in Fig. 3. Careful comparison of the two contour plots shows that, as expected, HSQC does offer superior resolution, i.e., the contour reponses are "tighter" or "sharper" than corresponding responses in the HMQC spectrum. Whether this benefit is significant or cosmetic cannot be readily assessed and will depend upon the individual problem being studied.

Beyond the higher resolution offered by the HSQC experiment, some other conclusions can be drawn from the experience of the authors in making this comparison. First, pulse calibration for the HSQC experiment is considerably less forgiving than for the HMQC experiment. As will be immediately noted on comparison of the HMQC and HSQC pulse se-

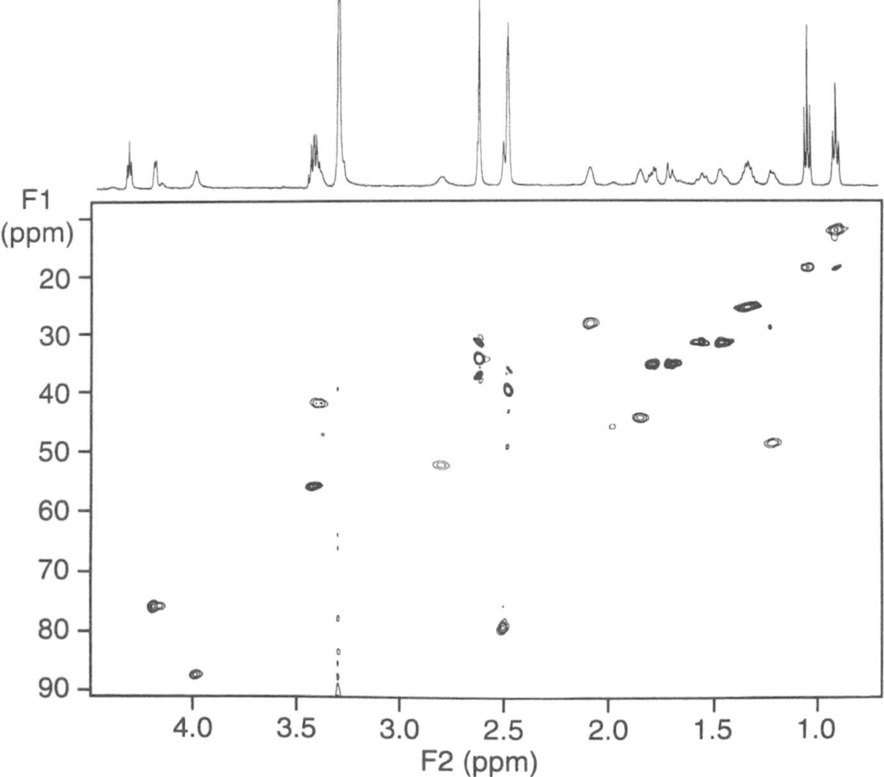

Fig. 4. The DEPT-HMQC (Kessler et al. 1989b) spectrum of the aliphatic region of ajmaline (**5**). The adjustable pulse (see Martin and Crouch 1991) was set $\beta = 180°$ to afford a spectrum in which the methylene resonances have positive intensity and the methine and methyl resonances are negative. Positive methylene responses are presented as *solid black contours*; negative methine and methyl responses are shown as *open contours*. Practically, the DEPT-HMQC spectrum shown accomplishes the same task as the recently reported GEM-COSY experiment (Domke et al. 1991) but, in our opinion, is more convenient than the GEM-COSY experiment

quences shown in Figs. 1 and 2, respectively, the former makes no use of 180° pulses on the X or heteronucleus channel of the instrument, while the latter has two. A miscalibrated 180° X-channel pulse in the HSQC experiment can significantly degrade the cancellation of unwanted $^1H-^{12}C$ components of magnetization. In contrast, the HMQC experiment is much more forgiving; pulses on the X-channel can be off by as much as 1–2 dB but will still allow the experiment to provide reasonable sensitivity with moderate samples. DEPT-HMQC uses one 180° X-pulse and also exhibits some sensitivity to the X-pulse calibration. The ability to edit spectra using DEPT-HMQC may, however, compensate for losses in spectral quality arising from poorly calibrated 180° X-pulses.

The edited HMQC spectra of a simple alkaloid recorded using the DEPT-HMQC experiment are presented in Fig. 4. The spectra were recorded using the condition $\beta = 180°$ to afford a spectrum with CH_2 signals with positive intensity and inverted CH and CH_3 responses. Some further interesting modifications of the DEPT-HMQC experiment are possible and will be discussed in the section below describing HMQC with isotropic mixing.

5 Long-Range Heteronuclear Chemical Shift Correlation – HMBC

To date, the only inverse-detected heteronuclear chemical shift correlation experiment available for multiple-bond correlation is the HMBC experiment of Bax and Summers (1986). The pulse sequence is shown in Fig. 5 and

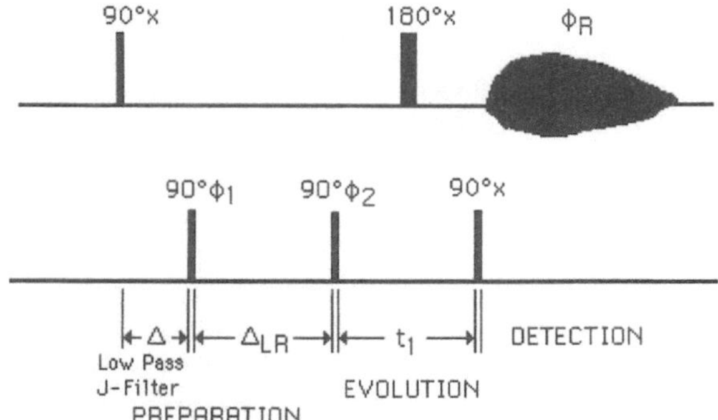

Fig. 5. HMBC pulse sequence of Bax and Summers (1986). The first 90° ^{13}C pulse serves as a low-pass J-filter (Kogler et al. 1983), as discussed in the text. Heteronuclear multiple-quantum coherence of order zero and two is created by the second 90° ^{13}C pulse. The 180° ^1H pulse interchanges zero- and double-quantum coherences and decouples proton chemical shift evolution during t_1. Observable proton single-quantum coherence is recreated by the final 90° ^{13}C pulse and detected

differs from the HMQC experiment (Fig. 1). The experiment begins with a pulse sequence element known as a low-pass J-filter (Kogler et al. 1983). Quite simply, the first 90° ^{1}H pulse and the 90° ^{13}C pulse following $1/2(^1J_{CH})$ later create heteronuclear multiple-quantum coherence for the direct or one-bond proton-carbon pairing. This, obviously, is an undesired component of magnetization in an experiment designed to observe long-range heteronuclear correlations. To eliminate one-bond responses, on alternate scans the phase of the first ^{13}C 90° pulse is inverted so that the signal from the one-bond correlation is alternately added and subtracted, effectively eliminating direct responses. During the period of time represented by the low-pass J-filter, the long-range component of magnetization evolves much more slowly and is essentially unaffected. Some 50 to 80 ms after the initial 90° proton pulse, the long-range component of magnetization is oriented for creation of multiple-quantum coherence by the second 90° ^{13}C pulse. Zero- and double-quantum coherences are created, evolve, are interconverted by the 180° proton pulse, and are reconverted to observable proton single-quantum coherence by the final 90° ^{13}C pulse. Observable ^{1}H signals arising from reconversion of heteronuclear multiple-quantum coherence are modulated by ^{13}C chemical shifts and homonuclear scalar couplings. Hence, HMBC spectra are often presented in an absolute value mode.

An alternative to absolute value presentation has been proposed by Bax and Marion (1988). We have found mixed-mode processing, in which the data is absorptive in the ^{13}C (F_1) frequency domain and absolute-value-calculated in the proton (F_2) frequency domain to provide a superior presentation of HMBC spectral data. Williamson et al. (1989) carried the idea of absorptive HMBC experiments a step further, demonstrating in the specific case of the antibiotic distamycin-A that fully phase-sensitive HMBC spectra may be recorded.

Selection of parameters and the processing of HMBC data have been considered in some detail in the recent review of Martin and Crouch (1991). Hence we will discuss this only very briefly here. Optimization of the long-range delay for about 50 ms (10 Hz) is a useful starting point for aromatic compounds; 80 ms works well for aralkyl molecules. We default the experiment to 63 ms, leaving any final adjustment to the whim of the individual user. Practically, we have found it useful to use an interpulse delay about 50% longer with the HMBC experiment than with the HMQC experiment on the same compound, defaulting the experiment to a delay of about 1.8 s. Obviously, a trade-off must be made between numbers of transients acquired and digitization in the F_1 frequency domain. Beyond the minimum number of transients required to complete a phase cycle, we generally have found it preferable to opt for increased F_1 digitization rather than an increased number of transients/t_1 increment. We also find it useful to increase the number of data points acquired in t_2 to about 4K points. Three-bond heteronuclear correlation responses are predominant in an HMBC spectrum. Two- and, less often, four-bond responses may also be observed, both

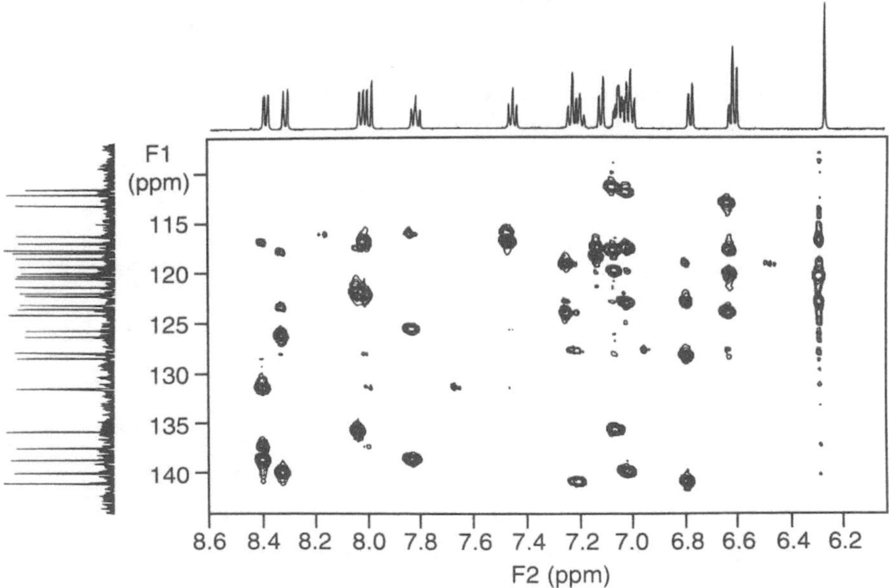

Fig. 6. HMBC spectrum of cryptospirolepine (**1**) in d$_6$-DMSO recorded at 400 MHz. While most of the responses contained in the spectrum arise via long-range heteronuclear coupling pathways of interest, there are several incompletely canceled direct responses still observable. Incompletely canceled direct responses in an HMBC spectrum are readily identified as a 120–180 Hz doublet centered about the proton chemical shift at the carbon chemical shift of the directly bound carbon identifiable from the HMQC spectrum

typically less intense than three-bond responses. Other than response intensity, there is unfortunately no way in which to differentiate three-bond from other correlation pathways in an HMBC spectrum. Generally, though, there will be sufficient redundancy in the data to make it obvious when a two- or four-bond correlation response is encountered.

HMBC spectra are extremely useful in their ability to unequivocally assign quaternary carbons and to link protonated carbon structural components to one another through intervening quaternary carbons or across heteroatoms. In this regard, HMBC data can be considered complementary to "through-space" correlation (NOESY/ROESY) data. Consider the structure of cryptospirolepine (**1**) (Tackie et al. 1993). During the elucidation of the structure, two of the four four-spin systems were unequivocally identified using a COSY spectrum. The other two four-spin systems required the use of HMQC-TOCSY (discussed below) for the complete and unequivocal identification of the individual constituents. Linking these structural fragments into larger contiguous substructures was accomplished using both HMBC and ROESY data. As illustrated by structural fragment **2** and Fig. 6, the protons resonating at 8.400 and 8.004 ppm were long-range correlated to quaternary carbons resonating at 138.5 and 116.8 ppm, respectively. The N-methyl resonating at 4.365 ppm was long-range coupled to quaternary carbons resonating at 138.5 and 123.4 ppm. The former long-range coupling of the N-methyl, in conjunction with an observed nOe to the proton of the four-spin system resonating at 8.004 ppm, established the location of these structural components relative to one another. Using this

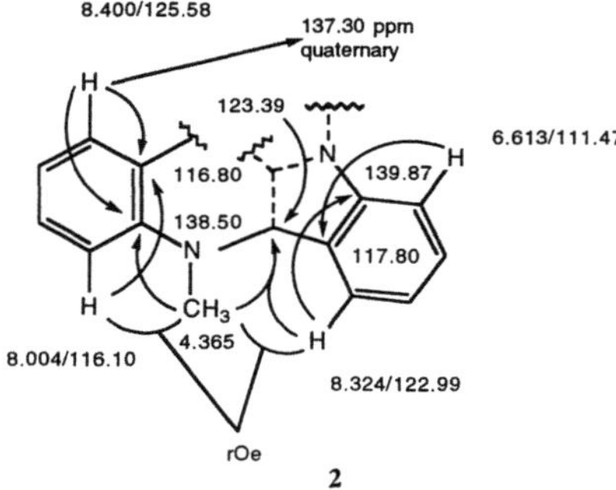

approach, the balance of the structural fragment represented by **2** was assembled.

There have been numerous applications of the HMBC experiment in the elucidation of alkaloid structures, which are discussed in the applications sections (see Sect. 9) of this chapter. One other application that bears mention at this point has yet to be applied to an alkaloid: the application of long-range correlations between "identical" atoms in a molecule exhibiting C_2 symmetry. An example of this type of application was recently reported for the C_2-symmetric tetrastilbene hopeaphenol. The reader faced with this type of problem is referred to the excellent treatment of Kawabata et al. (1992b). In similar fashion, it is also possible to address stereochemical issues of this type (Kawabata et al. 1992a). For a discussion of an example of this type of application, the reader is directed to the section of this chapter dealing with HMQC-NOESY/ROESY.

6 HMQC-TOCSY and Other Hyphenated Inverse-Detected 2D Experiments

Despite its significant utility, the COSY spectrum can provide frustratingly little information when the proton spectrum becomes highly congested. As an example, consider two protons in different parts of the molecule that have essentially the same chemical shift. Even though they are not mutually coupled, responses in the COSY spectrum correlating other protons to either of the two overlapping resonances become confused. Frequently, there is no way of knowing to which proton a particular correlation leads. In

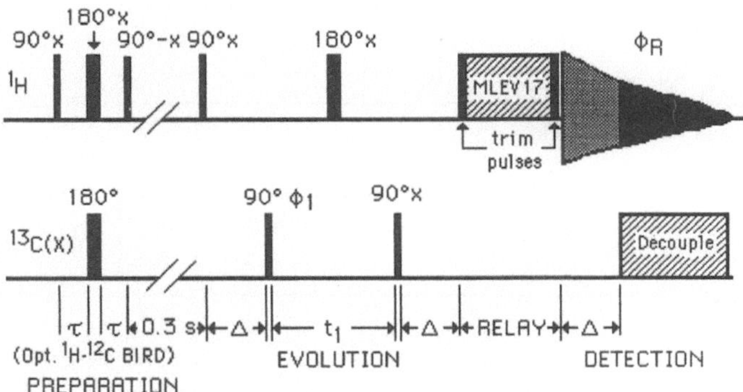

Fig. 7. HMQC-TOCSY pulse sequence described by Lerner and Bax (1986). Proton magnetization is manipulated and labeled with the chemical shift of the directly attached ^{13}C in a fashion identical to the HMQC experiment (see Fig. 1). After a refocusing period, Δ, proton magnetization is propagated from the directly attached proton to its vicinal neighbors using an MLEV-17-based isotropic mixing period. In the original work, the receiver was enabled and broadband heteronuclear decoupling initiated after a fixed delay, $\Delta = 1/2(^1J_{CH})$, to suppress the direct responses. Alternative considerations regarding direct responses in an HMQC-TOCSY spectrum are discussed in the text

the case of a structure elucidation problem, this situation clearly leads to a dead end: an alternative means must be sought to establish the structure.

Heteronucleus-detected heteronuclear relay experiments (Bolton 1982; Bolton and Bodenhausen 1982; Bax 1983a; Kessler et al. 1983) first addressed the problem just considered. Unfortunately, these experiments were difficult to optimize when a range of vicinal couplings was involved in a fashion similar to that of the relayed COSY or RCOSY experiment (Bax and Drobny 1985). Optimization problems have been circumvented in the homonuclear case by the development of an experiment variously known as TOCSY or HOHAHA (Braunschweiler and Ernst 1983; Davis and Bax 1985a,b). The first effort to combine isotropic mixing with sorting protons via heteronuclear shift was reported by Bax et al. (1985a) in a heteronucleus-detected experiment. The first attempted proton-detected heteronuclear relay experiment was reported by Bolton (1985).

Lerner and Bax (1986) reported a pair of experiments. The first, HMQC-COSY, is substantially less versatile than the HMQC-TOCSY experiment shown in Fig. 7. Functionally, the experiment is an HMQC experiment to which a "TOCSY" or isotropic mixing step has been appended. After observable proton single-quantum coherence is recreated, a fixed delay, $\Delta = 1/2(^1J_{CH})$, brings antiphase components of proton magnetization into phase. In-phase proton magnetization is spin-locked (mixed) to propagate magnetization from the directly bound proton to its vicinal neighbor(s). The experiment thus affords proton-proton connectivity information equivalent to a COSY spectrum, sorted by the chemical shift of the carbon directly bound to the proton from which magnetization was propagated.

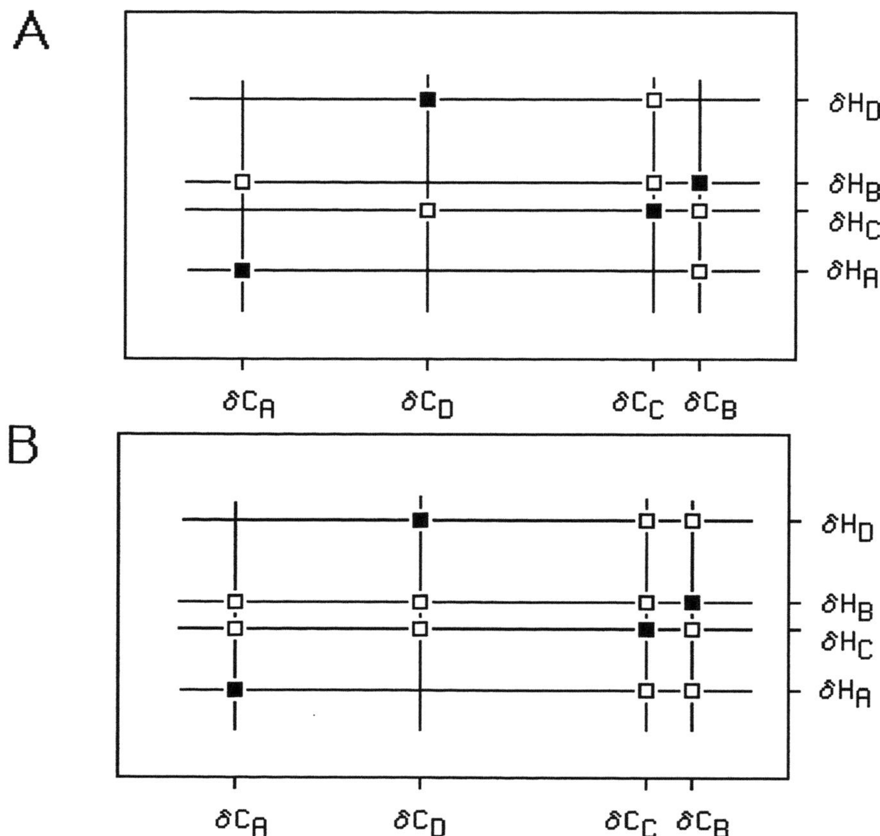

Fig. 8A,B. Schematic representation of results obtained from HMQC-TOCSY spectra of a 1,2-disubstituted aromatic system that would be expected for mixing times of about 18 to 24 ms (**A**) and 24 to 30 ms (**B**). The direct responses are denoted by *solid squares*, relayed responses are denoted by *open squares*. **A** During an 18- to 24-ms mixing period, magnetization would be propagated from H_A, directly bound to C_A to the vicinal neighbor proton, H_B. **B** During a 24- to 30-ms mixing period, magnetization would be propagated from H_A to H_B and then from H_B to H_C, giving responses along the axis defined by the chemical shift of C_A at the chemical shift of H_A (*solid square*) and both H_B and H_C (*open squares*)

6.1 Interpretation of HMQC-TOCSY Spectra

Unless a specially modified variant of the HMQC-TOCSY experiment is utilized (see below), interpretation of an HMQC-TOCSY spectrum requires prior knowledge of the location of the direct proton-carbon correlation responses. To illustrate the interpretation of HMQC-TOCSY data, it is useful to begin with a simple schematic illustration of the connectivity network associated with the proton four-spin system of a 1,2-disubstituted phenyl ring. As shown in Fig. 8A, we may begin with the direct response for the proton-carbon pair H_A-C_A that would be identified in the HMQC spectrum. With short isotropic mixing times (typically 12–18 ms would be the starting point for an aromatic system) we would see responses along the proton frequency axis, F_2, at the chemical shift of H_A and H_B sorted in the carbon frequency domain, F_1, by the chemical shift of C_A. (These correlations appear vertically along the axis defined by δC_A in Fig. 8A; we observe responses at δH_A and δH_B. The reader should realize that this axis could easily be reoriented orthogonally by transposition of the data matrix and that the orientation is immaterial because the information content is identical irrespective of orientation.) Conversely, at the chemical shift of C_B we would expect to see a response at the chemical shift of H_B and responses arising from the propagation of magnetization from H_B to both H_A and H_C. At longer mixing times, typically 24–30 ms for an aromatic system, additional responses are observed as illustrated schematically in Fig. 8B. As mixing time is increased further, propagation can be expected to continue to more distantly removed protons comprising the spin system.

Using the approach just described to interpret HMQC-TOCSY spectra is fine, provided, of course, that the resonances in the proton spectrum are resolved. This approach cannot be employed when there are significant overlaps in the data; correlations to the overlapped protons will lead to assignment and/or structural ambiguities. Worse still was the situation encountered with cryptospirolepine (**1**) (Tackie et al. 1992), in which the proton spectrum contained two overlapped AB portions of two ABXY spin systems. To illustrate the considerable utility of HMQC-TOCSY spectra, this problem will be examined in detail below. Before doing so, however, it is useful to consider an alternative means of extracting correlation information from the HMQC-TOCSY spectrum.

To this point, using the schematic presentations contained in Fig. 8, we have examined the spectrum for responses in the ^{1}H (F_2) frequency domain along the vertical axis defined by the chemical shift of the carbon directly bound to our starting proton, e.g., H_A. An attractive alternative method of interpreting the data exists. We may just as readily define an "interpretational axis" on the basis of a proton chemical shift, e.g., δH_A. Using this orthogonal frame of reference, we observe a direct response (HMQC) at the coordinates $\delta H_A/\delta C_A$, which will serve as our starting point. Examining the axis we have defined for other responses, we note a response at δC_B that arises from the relay of magnetization from H_B to H_A. Practically, the response defines in an indirect but useful manner, the chemical shift of C_B. Referring to Fig. 8B along the same "interpretational axis," we see the direct response at $\delta H_A/\delta C_A$, the single-relayed response at δC_B discussed above and, assuming that the mixing time is long enough for propagation to occur from $H_C \rightarrow H_B \rightarrow H_A$, a response at δC_C. Thus, indirectly, we have identified three contiguous protonated carbons in our hypothetical 1,2-disubstituted phenyl structural fragment. The utility of this approach was

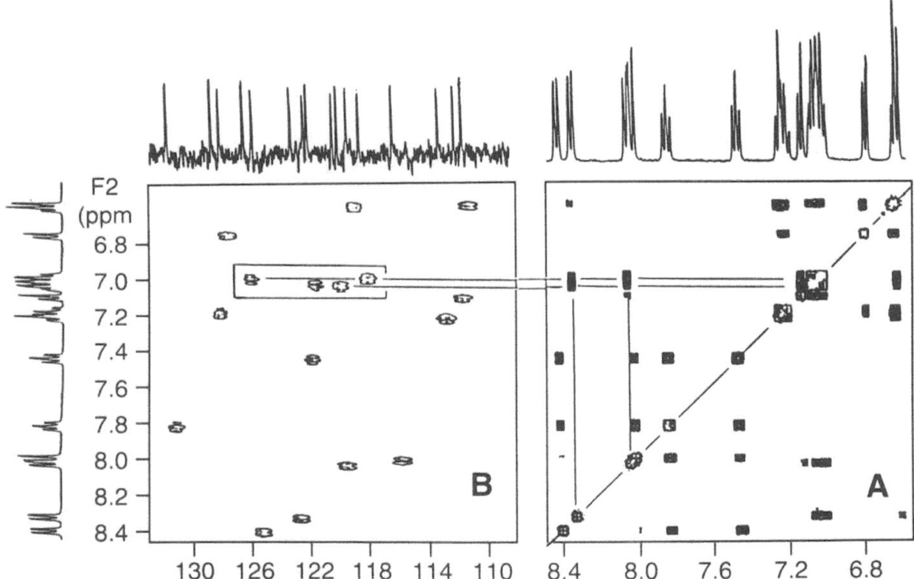

Fig. 9. A Portion of the COSY spectrum of cryptospirolepine. **B** Corresponding region of the HMQC spectrum of cryptospirolepine. The carbons directly bound to the AB protons of the two ABXY spin systems are enclosed within the *boxed region* in the HMQC spectrum

first demonstrated experimentally in the case of a simple degradation product of a sesquiterpene (Martin and Crouch 1991) and more recently using the alkaloid quindoline as a model compound (Spitzer et al. 1991).

A portion of the COSY spectrum of cryptospirolepine (**1**) is shown in Fig. 9 plotted beside the HMQC spectrum of the corresponding region of the proton spectrum. The two overlapped AB portions of the two ABXY spin systems are located in the center of the COSY spectrum. The involved protons have chemical shifts of 7.009, 7.012, 7.053, and 7.059 ppm. While the protons are heavily overlapped, the carbon resonances are, in contrast, quite well resolved, resonating at 118.35, 120.14, 121.86, and 126.19 ppm. The HMQC-TOCSY spectrum recorded with a mixing time of 18 ms is shown in Fig. 10. Beginning from the proton-carbon pair resonating at 8.324/122.99 ppm, we observe a relayed response to either the proton resonating at 7.053 or 7.059 ppm. Here we encounter the first point of ambiguity. The proton at 7.053 or 7.059 ppm is coupled, in turn, to either the proton at 7.009 or 7.012 ppm, creating a second level of ambiguity. Given these ambiguities, there are four assignment permutations possible and unavoidable from the COSY data.

In contrast, using the axis defined by the proton resonating at 8.324 ppm, we observe, in addition to the direct response at 8.324/122.99 ppm, a relayed

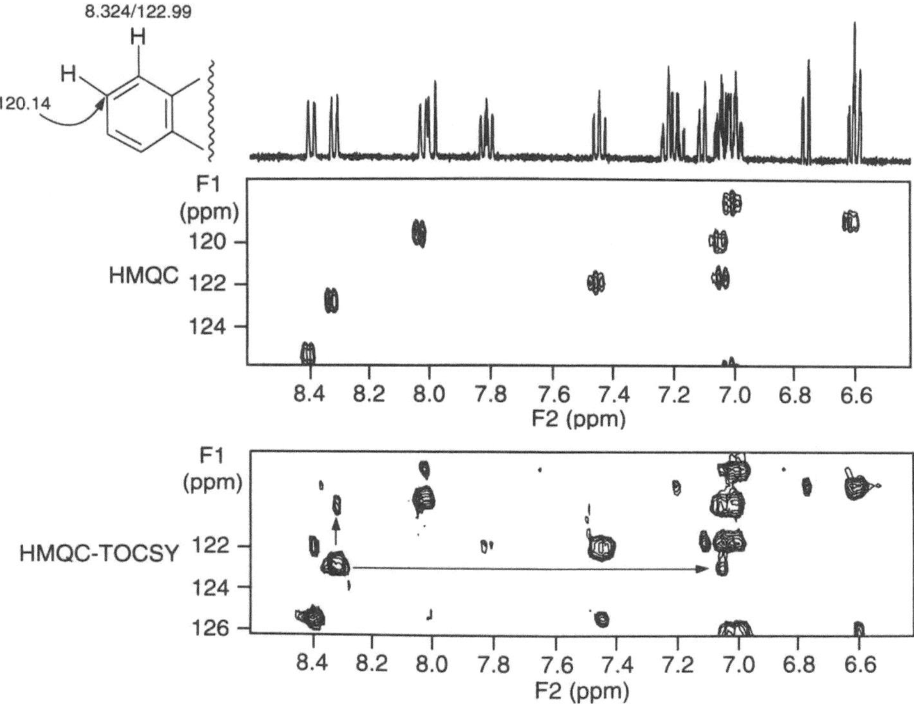

Fig. 10. HMQC-TOCSY spectrum of cryptospirolepine recorded using a mixing period of 18 ms. The proton-carbon pair resonating at 8.324/122.99 ppm discussed in the text exhibits a response (horizontally) to either the proton resonating at 7.053 or 7.059 ppm. The assignment of the vicinal neighbor is equivocal by this method, because the digital resolution in the F_2 frequency domain is too low to differentiate protons with a chemical shift difference of only 0.003 ppm. In contrast, we observe a response (vertically) at 120.14 ppm. This response identifies the protonated carbon bound to the proton vicinally coupled to the proton resonating at 8.324 ppm. From the HMQC spectrum, the vicinal proton is assigned at 7.059 ppm

response at 120.14 ppm. The proton directly bound to this carbon resonates at 7.059 ppm and is hence the vicinal neighbor of the proton resonating at 8.324 ppm. In a similar fashion, an HMQC-TOCSY spectrum recorded with a longer mixing time (not shown) establishes that the next contiguous protonated carbon is established as the resonance at 126.19 ppm. The correlation arises, as described above, from propagation 7.012→7.059→8.324 (H_C→H_B→H_A). Thus, despite being faced with four potentially interchangeable sets of assignments for this spin system, the assignments of all four of the protons and the directly bound carbons are unequivocally established using the HMQC-TOCSY experiment.

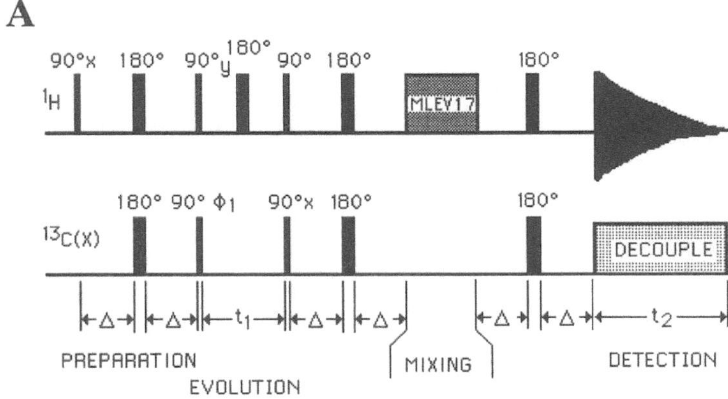

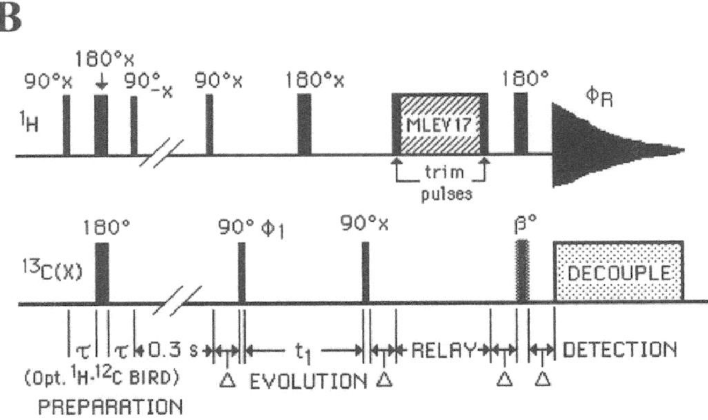

Fig. 11. A Pulse sequence to invert direct (one-bond) responses based on HSQC-TOCSY (Domke 1991). **B** Pulse sequence to invert direct responses based on HMQC-TOCSY (Martin et al. 1992). When the adjustable pulse, β, is set to 180°, the direct responses are inverted as in the original work of Domke (1991). In contrast, when $\beta = 90°$, direct responses will be canceled in a manner analogous to the procedure used to calibrate decoupler pulses (Thomas et al. 1981; Bax 1983b). In experiments when the direct response is to be canceled, there is no need of broadband heteronuclear decoupling during acquisition, allowing higher levels of digital resolution in F_2 than would otherwise be possible

6.2 Modifications of the HMQC-TOCSY Experiment – Inverted Direct Responses

There are a number of possible variants and modifications of the HMQC-TOCSY experiment. One obvious modification would be to use HSQC rather than HMQC to label protons with the chemical shift of the carbon to which they are directly bound. To the best of our knowledge, applications of

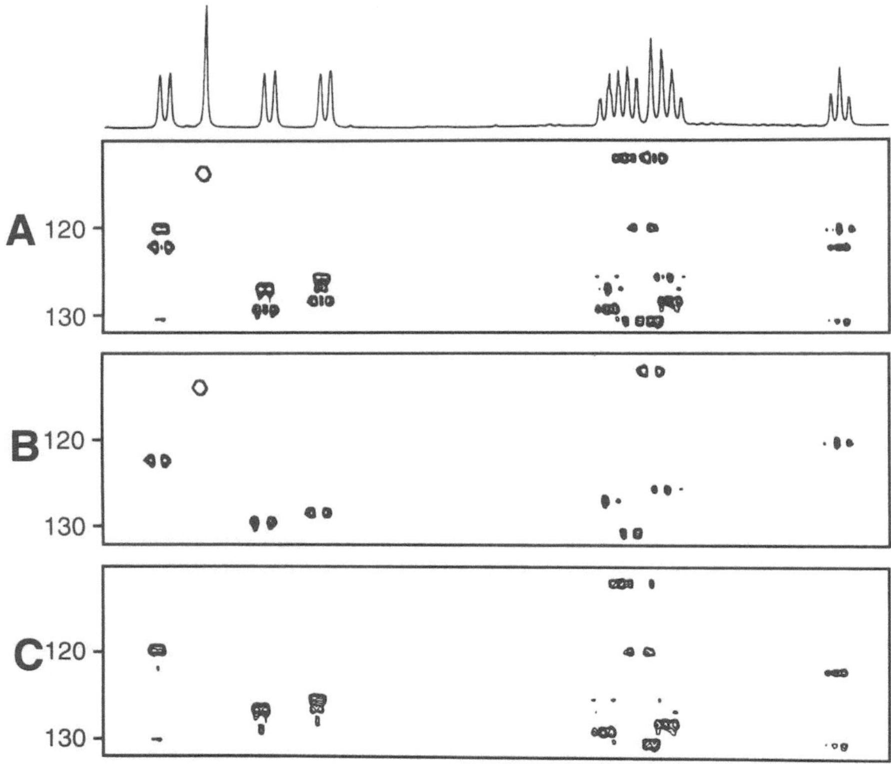

Fig. 12A–C. HMQC-TOCSY contour plots recorded for the alkaloid quindoline (**3**) using the pulse sequence shown in Fig. 11B. The spectrum was recorded with a mixing time of 24 ms. **A** Contour plot showing all responses. **B** Contour plot in which only negative contours from inverted direct responses are plotted. **C** Contour plot in which only the positive TOCSY responses appear

this particular experiment have not been reported. A modification of the HSQC-TOCSY experiment has, however, been described by Domke (1991). One of the problems inherent to HSQC-TOCSY or HMQC-TOCSY is the identification of the direct response. Domke has addressed this problem by applying a pair of simultaneous 180° ^{1}H/^{13}C pulses $\Delta = 1/2(^1J_{CH})$ after the isotropic (TOCSY) mixing period. After the pulses, a second delay, $\Delta = 1/2(^1J_{CH})$, is allowed to elapse before acquisition and broadband heteronuclear decoupling are initiated. This modification inverts the direct response making them identifiable, potentially obviating the need to acquire an HSQC or HMQC spectrum if HSQC- or HMQC-TOCSY are to be employed in structure elucidation. The pulse sequences for HSQC- and HMQC-TOCSY are shown in Fig. 11A and B, respectively. Although we have not specifically evaluated the former version, the latter works reasonably well, as illustrated using quindoline (**3**) (Fig. 12).

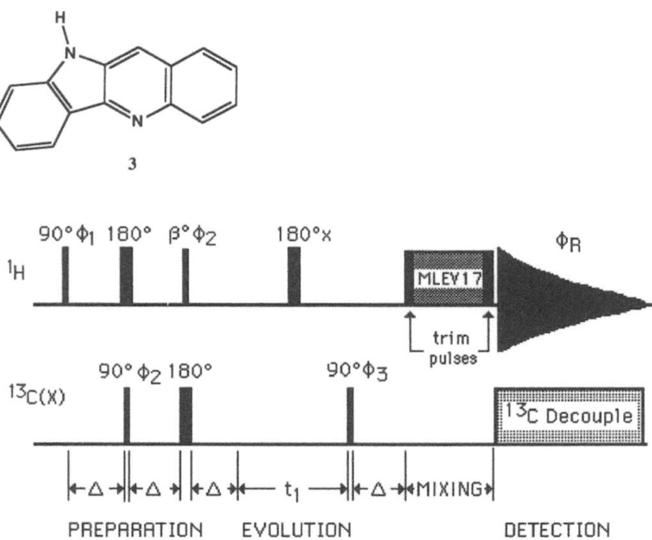

Fig. 13. DEPT-HMQC-TOCSY pulse sequence (Crouch et al. 1992b) originally suggested by Kessler et al. (1989b) but never developed in those laboratories. The editing function of the original DEPT-HMQC experiment is retained, the phase selected by the beta pulse; with $\beta = 90°$ only a methylene spectrum is obtained. More usefully, with $\beta = 180°$ a spectrum is obtained in which the direct and TOCSY signals originating from methylene groups have positive intensity and those originating from methine and methyl groups have negative intensity

6.3 DEPT-HMQC-TOCSY

The capability to further edit HMQC-TOCSY-type spectra was suggested by Kessler et al. (1989b) in the form of a DEPT-HMQC-TOCSY experiment. We are aware of only a single reported application of this experiment to analogs of cyclosporin (Seebach et al. 1991). Recent work in these laboratories (Crouch et al. 1992b), using the DEPT-HMQC-TOCSY pulse sequence, shown in Fig. 13, has firmly underscored the viability of the technique. By performing a single DEPT-HMQC-TOCSY experiment with

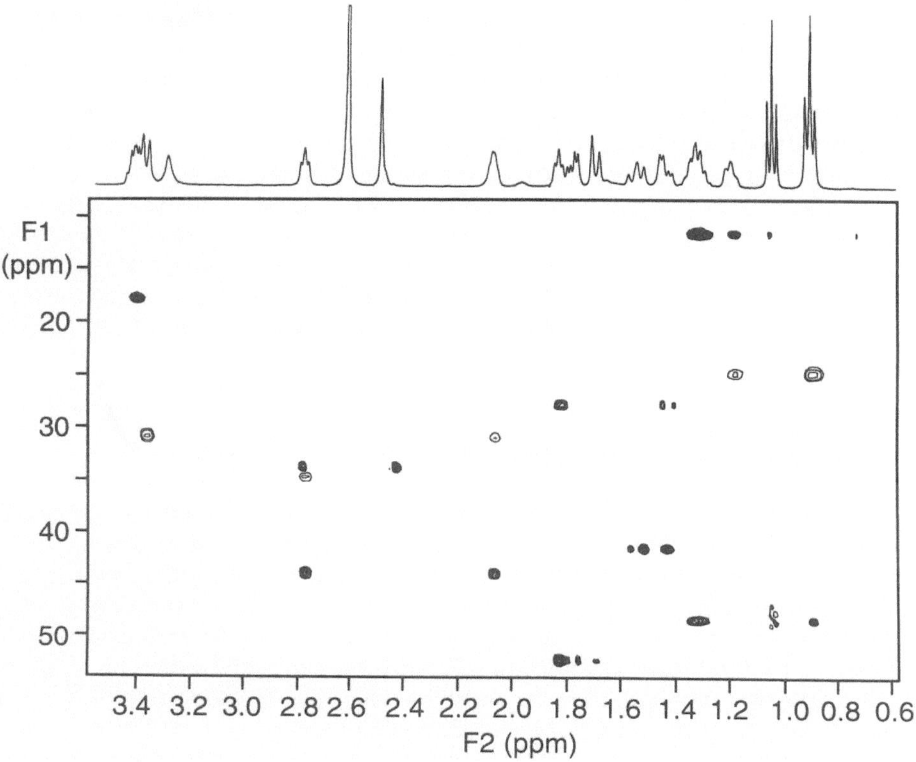

Fig. 14. Edited DEPT-HMQC-TOCSY spectra of a ajmaline (**4**). The spectrum was acquired using the pulse sequence shown in Fig. 13 (Seebach et al. 1991) modified to incorporate a 180°/90° $^1H^{13}C$ pulse pair located as shown by the pulse sequence in Fig. 11B. Direct ($^1J_{CH}$) responses were suppressed by the 180°/90° pulse pair (Crouch et al. 1992b; Martin et al. 1992). Responses arising through a relay from protons on a methylene carbon appear with positive intensity and are denoted in this presentation by the *solid responses*. Relayed signals arising through transfer from protons on a methine or methyl carbon appear with negative intensity and are denoted by the *open contours*

the pulse $\beta = 180°$, a phase-edited TOCSY spectrum is obtained in which responses originating from methine and methyl protons appear with negative intensity (inverted) and those from methylene protons are observed with positive intensity.

The DEPT-edited HMQC-TOCSY spectrum of ajmaline (**4**) is illustrated in Fig. 14. The spectrum was acquired using the pulse sequence shown in Fig. 13 modified to include a 180°/90° $^1H/^{13}C$ pulse pair position in a fashion analogous to that in Fig. 11B. The pulse pair serves to suppress unwanted direct responses (Crouch et al. 1992b; Martin et al. 1992). Relayed responses shown in Fig. 14 originating from protons on methylene carbons are observed with positive intensity and appear as solid black responses.

Responses originating through relay from protons on methine or methyl carbons appear with negative intensity and are denoted in Fig. 14 by the open contours. It is quite probable that DEPT-HMQC-TOCSY, in conjunction with DEPT-HMQC, may eventually become methods of choice in dealing with aralkyl alkaloids. These methods provide a convenient means of establishing carbon resonance multiplicities at high sensitivity. Further, the edited TOCSY experiment also provides a potential means of disentangling congested relayed responses in the aliphatic region of alkaloid spectra. In contrast, for molecules such as cryptospirolepine (1), where there is nothing to be gained by using the DEPT-edited variants of these experiments, HMQC and HMQC-TOCSY will continue to be the methods of choice.

6.4 Suppression of Direct Responses in HMQC-TOCSY Spectra

The original report of Lerner and Bax (1986) afforded a means to suppress direct responses in HMQC-TOCSY spectra that relied on the timing of the initiation of broadband heteronuclear decoupling relative to the beginning of the acquisition period. As shown in Fig. 7, by delaying the start of broadband heteronuclear decoupling by $\Delta = 1/2(^1J_{CH})$, the one-bond component of proton magnetization will become antiphase and hence eliminated by initiating decoupling. Although certainly workable, the procedure initially suggested by Lerner and Bax does intrinsically limit digital resolution in F_2 since decoupler duty cycles must still be considered. In many cases, limitations imposed by decoupler duty cycles will not be a concern. However, when dwell times correspondingly increase with decreasing F_2 spectral widths, the available F_2 digital resolution may be less than desirable. An example where this would become a problem would be a fully aromatic alkaloid.

A convenient way of circumventing the limitations imposed by heteronuclear broadband decoupling is to eliminate the direct responses completely by a means that does not require decoupling. At first thought, this might seem difficult, but it is really quite simply accomplished. Recall that decoupler pulses can be calibrated by applying a 90° pulse with the transmitter, followed $1/2(^1J_{CH})$ later by a 90° pulse from the decoupler (Thomas et al. 1981; Bax 1983b). Referring to Fig. 11B we may accomplish this using the HMQC-TOCSY pulse sequence shown in which the adjustable pulse, β, is set to 90°. Rather than inverting the direct response as in Domke's (1991) experiment where β = 180°, the direct responses are instead canceled. Cancellation is afforded by converting the magnetization component for $^1J_{CH}$ to unobservable multiple-quantum coherence. Divergence of the optimized value of the $1/2(^1J_{CH})$ delay and the actual one-bond coupling will reduce the efficiency of the cancellation, as will miscalibration of the 90° heteronucleus pulse.

7 Other Inverse-Detected NMR Experiments – Selective and F_1 Region-Selected Experiments

As structure elucidation studies are conducted, questions will often arise that require, for example, only one piece of connectivity information from an HMBC or HMQC-TOCSY experiment. Some researchers will undoubtedly be forced to forego the acquisition of a particular spectrum for lack of or cost of instrument time. These situations can now be approached through the use of selective one-dimensional analogs of the two-dimensional NMR experiments described above. Although one can envision many such experiments, one group in particular, selective one-dimensional analogs of the HMBC experiment, which have been given the acronym SIMBA (Selective Inverse Multiple Bond Analysis), hold the promise of being particularly useful (Crouch and Martin 1991; Keniry and Poulton 1991). SIMBA and other selective 1D analogs of 2D experiments are described in the following sections.

In other cases, there may often be a need for relatively high digital resolution in two or more regions of the spectrum separated by regions that contain no information. Ideally, we would like to be able to acquire regionally selected spectra with levels of digital resolution in each commensurate with the information density. In the past, this has been difficult since there is, of course, no way of filtering in the second frequency domain. This problem is conveniently and elegantly overcome, however, through the use of selective pulses applied to the heteronucleus. Selective pulses have recently been reviewed by Kessler et al. (1991c); we have also recently comparatively evaluated several selective pulses in F_1 region-selected HMBC experiments (Crouch et al. 1992a).

7.1 Selective One-Dimensional Analogs of HMBC – SIMBA

Long-range heteronuclear connectivity information is often the key to solving a structural problem. In many instances, one or two pivotal long-range correlations are all that is needed to differentiate between structural possibilities. Until recently, it was necessary to acquire a full HMBC spectrum to obtain the few connectivities that might really be needed. This is no longer a problem since the introduction of selective one-dimensional analogs of HMBC.

Berger (1989) anticipated the solution to the problem of needing experimental access to selective heteronuclear shift connectivities with the development of a selective 1D analog of the HMQC experiment to which he gave the acronym SELINCOR. Simply, SELINCOR replaces the final 90° carbon pulse of the HMQC experiment (see Fig. 1) with a selective Gaussian 90° pulse applied to the ^{13}C resonance of interest. SELINCOR has had no practical applications but was certainly of interest in a developmental sense.

The first experimentally viable selective 1D analog of HMBC appeared in the report of Keniry and Poulton (1991). Their experiment used a 90° proton pulse followed $1/2(^3J_{CH})$ later by a 90° ^{13}C pulse that created heteronuclear multiple-quantum coherence for protons long-range coupled to ^{13}C, nominally via three bonds ($^3J_{CH}$). Subsequently, a frequency-selective Gaussian pulse was applied to the carbon of interest, affording a proton spectrum composed of those resonances long-range coupled to the carbon to which the selective pulse was applied. Keniry and Poulton (1991) applied their "soft" HMBC experiment to lambertellin, a poorly soluble metabolite of the fungus *Ciboria gordonii*.

Several months after the work of Keniry and Poulton (1991), the SIMBA experiment was reported (Crouch and Martin 1991). SIMBA differed from the earlier report in that it was based on the HMBC pulse sequence (Bax and Summers 1986; see also Fig. 5). The SIMBA experiment retained the low-pass J-filter (Kogler et al. 1983) and the 180° proton pulse midway through a fixed duration 100 µs "evolution" period. As intially reported, the SIMBA experiment utilized a selective 90° square pulse in place of the last 90° ^{13}C pulse of the HMBC pulse sequence. During the course of further refinement of the SIMBA experiment, we noted that improved spectral results could be obtained by compensating for the duration of the selective pulse with a fixed delay of about 75% of the length of the selective pulse. Using a Gaussian 90° pulse significantly improved the selectivity of the experiment and reduced, somewhat, the duration of the fixed delay following the selective pulse. Utilizing a self-compensating Gaussian 270° (Kessler et al. 1990) pulse obviated the need for a fixed delay following the pulse. The most reliable variant of the experiment uses the E-BURP-2 pulse of Geen and Freeman (1991; see below).

The initial demonstration of the SIMBA experiment utilized the simple alkaloid norharmane (**5**) as a model compound. Comparison spectra showing

5

the HMBC spectrum, traces corresponding to the quaternary carbons from the HMBC spectrum, and SIMBA traces are shown in Fig. 15. Another interesting feature of the experiment is that the proton resonances observed, which are long-range coupled to the carbon to which the selective pulse is applied, appear as the normal proton multiplet with one additional splitting. The additional coupling information contained in the proton multiplet in a SIMBA experiment arises from the heteronuclear coupling between the selectively pulsed carbon and the proton in question. If coupling constants

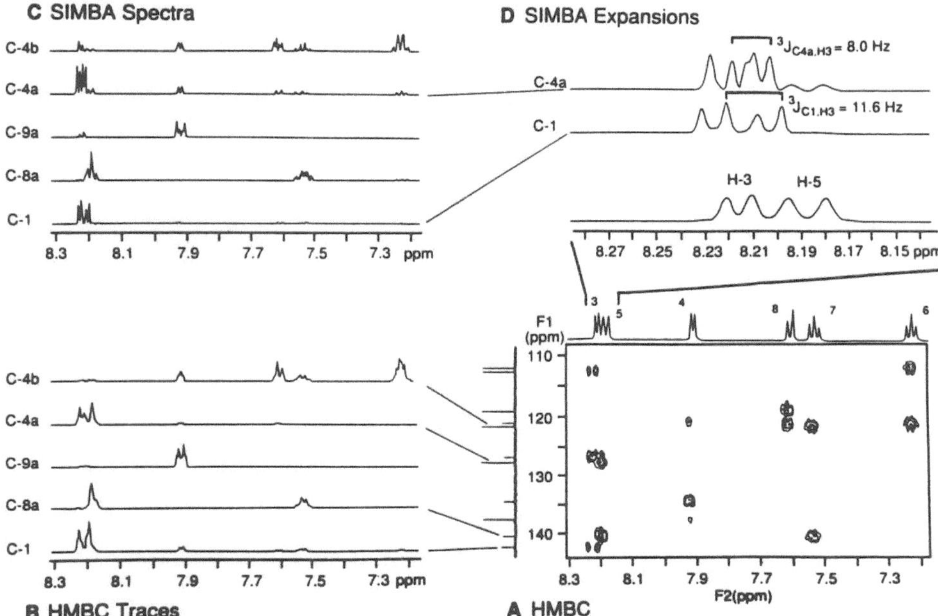

Fig. 15A–D. Comparisons of the HMBC and SIMBA spectra for the simple model alkaloid harmane (**5**). Panel **A** shows the HMBC spectrum of an 0.09 M solution in d_6-DMSO; acquisition time was about 2 h. Panel **B** shows the traces from the HMBC spectrum showing proton responses to each of the quaternary carbon resonances. Panel **C** shows the SIMBA traces for the same quaternary carbons. Some of the additional weak responses in the traces in this panel were a result of our initial attempt to illustrate the technique using square pulses, which can be implemented on a broader range of instruments than the more recent shaped pulses. Panel **D** shows, in more detail, the proton multiplets for C4a and C1, illustrating the appearance of the long-range heteronuclear coupling superimposed over the normal proton multiplet structure (used with the permission of Academic Press)

to a particular carbon represent useful structural information, the SIMBA experiment provides an unequivocal means of obtaining this information. One of the panels shown in Fig. 15 also illustrates the appearance of the additional splittings of two of the resonances relative to a normal proton reference spectrum.

As noted above, one implicit impediment to the utilization of SIMBA is the requirement of an accurate ^{13}C chemical shift for the resonance to be pulsed selectively. The problem is exacerbated rather than ameliorated when Gaussian 270° pulses are employed. The need for an accurate ^{13}C chemical shift is circumvented by utilizing the recently reported E-BURP-2 pulse of Geen and Freeman (1991) for the final 90° ^{13}C pulse of the SIMBA experiment. Since the E-BURP-2 pulse gives uniform phase and excitation across a relatively wide bandwidth, in contrast to the Gaussian 270° pulse which provides uniform phase and excitation across only a relative narrow

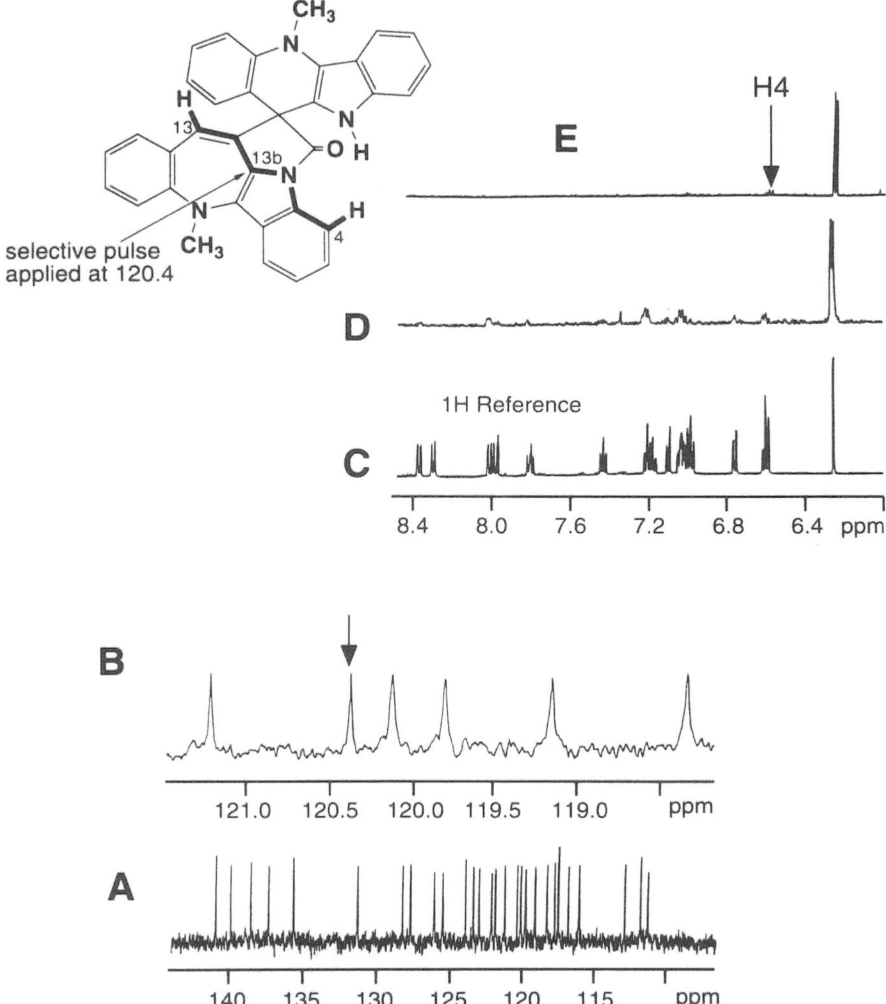

Fig. 16A–E. Application of the SIMBA experiment in the elucidation of the structure of cryptospirolepine (1). **A** Shows the aromatic region of the reference carbon spectrum. **B** Expansion of the aromatic region showing resonances in proximity to the resonance at 120.40 ppm (denoted by *arrow*) to which the selective pulse in the SIMBA experiment will be applied. **C** Proton reference spectrum. **D** Shows the SIMBA spectrum recorded with an 8-ms square pulse applied to C13b resonating at 120.4 ppm; the resonances flanking C13b in the carbon spectrum are the C6 resonance at 120.14 ppm and the C5a' resonance at 121.21 ppm. Some of the extraneous responses in this trace undoubtedly arise from the partial excitation of C6 and/or C5a' when a square pulse is used. **E** SIMBA spectrum acquired using a 12-ms 90° Gaussian pulse to excite C13b. The only responses observed are those correlating C13b to H13 and H4 as shown by the connectivity pathways. As demonstrated in this figure, the results obtained using a 90° Gaussian pulse are far superior to those obtained with a square pulse. Results obtainable with a Gaussian 270° pulse are better still than those with a Gaussian 90°

region, the E-BURP-2 variant of the SIMBA experiment can be used effectively even when only an approximate (± 0.5–1.0 ppm) chemical shift for the carbon resonance is known, e.g., from a response in an HMBC spectrum.

Applications of the SIMBA experiment in the literature thus far have been relatively limited in number. Keniry and Poulton (1991) used "soft" HMBC to assign the quaternary carbon resonances of lambertellin, a pyranonaphthaquinone fungal metabolite. More recently, we have demonstrated the use of the SIMBA experiment to confirm connectivities to the quaternary carbon resonances of the simple indoloquinoline alkaloid cryptolepine (**13**), when the F_1 window was inadvertently set leaving one of the quaternary carbons on the very edge of the spectrum (Tackie et al. 1991). In a more interesting application, a SIMBA experiment was utilized in assembling the structure of the complex alkaloid cryptospirolepine (**1**) (Tackie et al. 1993). Despite having an excellent HMBC spectrum of the molecule, connectivities from the quaternary carbon resonating at 120.4 ppm, ultimately assigned as C13b, did allow the construction of the indolobenzazepine. C13b exhibited, in the HMBC spectrum, only a correlation to the H13 vinyl resonance (via $^3J_{CH}$) of the benzazepine system. A SIMBA spectrum acquired in which the selective pulse was applied to the C13b carbon resonating at 120.4 ppm, shown in Fig. 16, did provide new structural information. Specifically, when a 12-ms Gaussian 90° pulse was employed, in addition to the long-range correlation to the H13 vinyl proton of the benzazepine, an additional weak correlation was observed to the proton resonating at 6.613 ppm which was ultimately assigned as H4. The $^4J_{CH}$ correlation pathway in this case is shown by **6**. For purposes of comparison,

6

also shown as the trace in Fig. 16D is the SIMBA spectrum acquired using a square excitation pulse. Undesired responses observed in the trace, of comparable intensity to the information sought and shown in the top trace, arise from the inferior selectivity of the square pulse relative to the 90° Gaussian pulse.

The application of SIMBA to cryptospirolepine (**1**) highlights another very important usage of selective one-dimensional experiments. While it was not possible to observe the connectivity between C13b–H4 in the full HMBC spectrum, probably due to insufficient signal-to-noise, responses of this type can be sought in a time-efficient manner using selective one-dimensional experiments! Sufficient numbers of transients can be accumulated to attain virtually any desired signal-to-noise level without having to invest the time to do this for numerous increments of the evolution period, t_1. Alternatively, spectra for a number of optimizations of the long-range delay could be acquired and examined. In our opinion, however, the former would be the normal approach; varying the optimization of the long-range delay in HMBC spectra has relatively little effect on the responses observed in the spectrum (Martin and Crouch 1991).

7.2 One-Dimensional HMQC-TOCSY

An experiment that actually preceded SIMBA, but which probably has somewhat lower utility, was the selective 1D HMQC-TOCSY experiment described by the authors (Crouch et al. 1990c). The selective experiment differs from the HMQC-TOCSY pulse sequence only in the replacement of the final ^{13}C 90° pulse by a selective pulse. Initially, we utilized a square pulse in this application but, obviously, the improvement in the SIMBA experiment by using a Gaussian 270° or an E-BURP-2 pulse would be realized in this experiment as well. To the best of our knowledge, there have been no applications of this experiment to alkaloids reported in the literature.

7.3 F$_1$ Region-Selected HMBC

One problem inherent to the HMBC experiment is attaining sufficient digital resolution when very wide F_1 spectral widths must be employed to include quaternary aromatic or carbonyl carbons. The same problem occurs with aralkyl alkaloids. Frequently, there may be a number of aromatic carbons separated by a wide region containing few or no responses from a congested aliphatic region.

Electing to restrict the F_1 spectral width is inappropriate, since responses to carbon resonances outside of the window can fold and could be mistaken for responses to a given proton if the fold is unfortuitous. This concern is particularly valid when we consider that in some cases it is easier to obtain ^{13}C chemical shift data from inverse-detected experiments than to acquire a carbon NMR spectrum directly if sample size is severely restricted. The solution to this problem is again found in the use of selective pulses; pulses are chosen to uniformly excite a particular F_1 region rather than a single resonance as in the case of the SIMBA experiment.

A number of F_1 region-selected inverse-detected 2D NMR experiments have been reported. One of the earliest (Davis 1989) took a different approach from more recent reports. Recognizing the need for higher digital resolution in assigning peptide spectra, Davis approached F_1 region selection by limiting the protons excited during the experiment. Specifically, he applied "soft" proton 90° and 180° pulses using the HMBC pulse sequence (Bax and Summers 1986) shown in Fig. 5. In this fashion, "hard" carbon pulses may be used with impunity since only those carbons long-range coupled to excited protons are detected.

Following closely after Davis' report was a paper by Bermel et al. (1989) that utilized selective 90° Gaussian pulses applied to ^{13}C. While the approach worked, as noted subsequently by Kessler et al. (1990), part of the resolution gain is lost when 90° Gaussian pulses are utilized because phasing of the peaks to pure absorption lineshape is no longer possible. As noted by Kessler, several alternatives are available, including: a rising half-Gaussian pulse (Friedrich et al. 1987); a rising half-Gaussian pulse with a hard 90° purging pulse (Kessler et al. 1989a); and a Gaussian 270° pulse (Emsley and Bodenhausen 1989). Kessler et al. (1990) have advocated and successfully utilized HMBC-type experiments in which a Gaussian 270° pulse is employed as the first ^{13}C 90° pulse, with the low-pass J-filtering pulse omitted in their experiments. The Gaussian 270° pulse has also been used in the assignment of a number of peptides by Kessler and colleagues (Kessler et al. 1991a,b; Seebach et al. 1991). Following the report of Geen and Freeman (1991) describing E-BURP and related band-selective pulses, we have conducted a comparative evaluation of four possible variants of the HMBC experiment using both an E-BURP-2 and Gaussian 270° pulse (Crouch et al. 1992a). In our experience, superior results were obtained using the E-BURP-2 pulse as demonstrated using the simple alkaloid quindoline (**3**) and the cyclic immunosuppressive peptide cyclosporin-A.

7.4 HMQC-NOESY and HMQC-ROESY

Two additional inverse-detected NMR experiments have been reported: HMQC-NOESY (Sohn and Opella 1989) and HMQC-ROESY (Kawabata et al. 1992a). Although these experiments provide attractive capabilities, at natural abundance of the heteronuclide, they unfortunately also suffer from hideously lower sensitivity than the other inverse-detected experiments that we have discussed. Quite simply, the reason for our impeachment of these techniques is the dynamic range requirements they invoke. Observing the direct response in either experiment is simple. Unfortunately, signal-to-noise levels must be sufficient to allow the observation of the nOe or rOe response, that may only be a few percent of the intensity of the direct response. Consequently, applications of HMQC-NOESY or HMQC-ROESY will be restricted to those problems where the use of the experiment is inescapable (Crouch et al. 1990a,b; Kawabata et al. 1992a). Pulse sequences for HMQC-NOESY and HMQC-ROESY are shown in Fig. 17.

Perhaps the one way in which HMQC-NOESY/ROESY will find practical application will be when it is necessary to extract critical n/rOe connectivities that cannot be accessed in any other way. One such example has been reported for the C_2-symmetric tetrastilbene hopeaphenol in which it

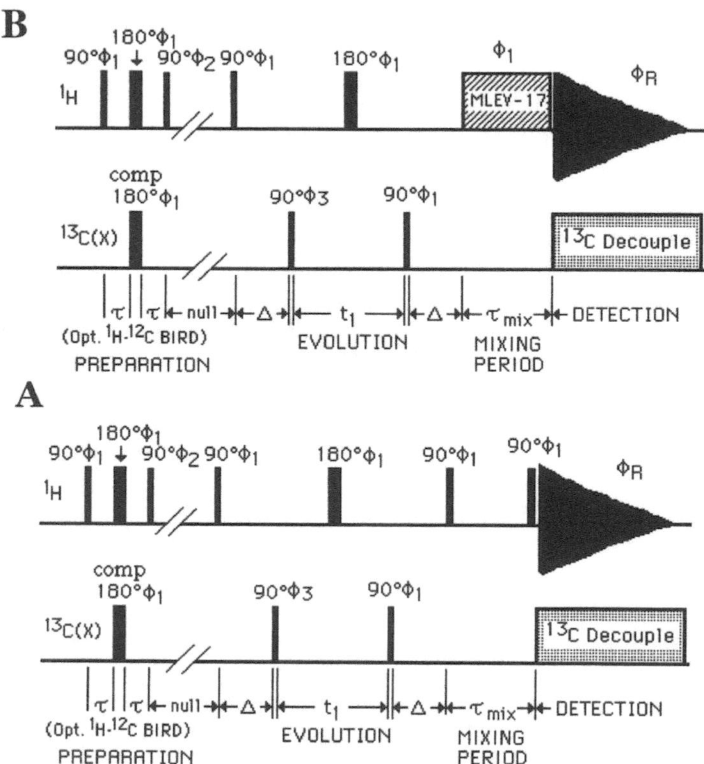

Fig. 17A,B. Pulse sequences for HMQC-NOESY (A) (Sohn and Opella 1989) and HMQC-ROESY (B) (Kawabata et al. 1992a)

was of interest to determine whether the two symmetrical components were *cis* or *trans* to one another. In this case, it was necessary to establish an rOe between two identical protons, with identical chemical shifts (Kawabata et al. 1992a). Although seemingly impossible, by recording an HMQC-ROESY spectrum without decoupling during the acquisition period, the two halves of the molecule may be differentiated isotopomerically. One of the two "identical" protons of interest will be bound to ^{13}C, the other to ^{12}C. When the spectrum is recorded without decoupling, the $^{13}C-^{1}H$ isotopomer will appear as a 125–180 Hz doublet due to $^{1}J_{CH}$. In contrast, the $^{12}C-^{1}H$ isotopomer will appear at the chemical shift. In the case of a molecule with C_2 symmetry in which the two halves are oriented such that the protons are *cis* to one another, an nOe or rOe response will appear in the center of the doublet arising from the $^{13}C-^{1}H$ isotopomer.

The one other way in which HMQC-NOESY/ROESY may see a wider range of application will be with the development of 1D analogs of these experiments. Neither has been reported as yet, but either could easily be performed given the provocation provided by a problem that lends itself to one of the techniques. In a one-dimensional selective analog of HMQC-NOESY/ROESY, the problem of lower sensitivity is ameliorated because only a single experiment is to be performed, as with the SIMBA technique. Signal-to-noise levels adequate to observe the needed nOe/rOe response, selected by the ^{13}C resonance selectively pulsed, can be obtained in a few hours rather than in perhaps the 24 h it might take to record the corresponding 2D spectrum.

8 Sample Requirements

Interestingly, between the first writing of this chapter and the time it was edited, there has been a substantial downsizing in the minimum sample on which inverse-detected heteronuclear chemical shift correlation may be conveniently acquired. Nominally, in the author's laboratories, HMQC spectra can be acquired in 1 h or less on about 0.5–1 mg of a medium-sized (300–500 Da) molecule at 500 MHz. HMBC spectra at this sample level may take from about 4 h to overnight depending upon the digital resolution in F_1 and the level of signal-to-noise sought. HMQC-TOCSY spectra for a 0.5–1 mg sample will typically consume 2–4 h. The corresponding DEPT-based experiments will require approximately twice these times. These general guidelines are based on the assumption of a contemporary 5-mm inverse detection probe.

As the available sample or the limitations of solubility begin to intrude, the time required to acquire a given spectrum, of course, increases. Practically, we find the lower limit for the acquisition of an HMQC spectrum overnight in a 5-mm probe to be about 80–100 µg. For the study of very minor alkaloidal constituents, degradation products, metabolites, etc., however, there may be occasions necessitating the acquisition of inverse-detected spectra at still lower concentrations. While not practical using a 5-mm probe, it is possible to acquire HMQC spectra on samples approaching 10 µg by using a micro inverse-detection probe. Recently, for example, we have reported an HMQC spectrum on 12 µg of cryptolepine (**13**) overnight (16 h); an HMBC spectrum was acquired on 35 µg of cryptolepine (**13**) in 21 h (Crouch and Martin 1992a). A comparative evaluation of a 5-mm probe with a micro inverse probe (Crouch and Martin 1992b) has demonstrated a factor of two greater signal-to-noise for the micro inverse probe relative to the conventional 5-mm probe. Practically, this translates to a factor of four reduction in the length of time required to reach a given signal-to-noise level. Further refinements in micro inverse probe designs may push these levels still lower. Comparable advances have also been experienced with the acquisition of ^{13}C reference spectra using either micro dual or carbon-optimized micro probes.

9 Applications of Inverse-Detected NMR Techniques to Alkaloid Chemistry. Structural Problems

In attempting to survey applications of inverse-detected 2D NMR techniques in alkaloid chemistry, we have made every effort to be comprehensive. However, the accuracy of any review can only be as good as the referencing done by the authors using various NMR techniques. As techniques progress from new and novel to accepted and routinely used, authors unfortunately tend to stop citing the primary reference to the technique in question. At this point, unless the reviewer is willing to sit and laboriously pare through the myriad of journals in which applications papers may appear in conjunction with computer searches, it is probable that some excellent applications will be missed in the process. In this regard, we wish to apologize to any authors whose applications of inverse-detected NMR techniques to alkaloid chemical problems we may have missed in the process of attempting to review the field.

In examining the applications of inverse-detected 2D NMR techniques that we have uncovered, the experiments have been applied to a very wide range of alkaloid types. Applications discussed cover the period from 1986, when inverse-detected techniques first began to be utilized through early 1992. Practically, there were no applications of even HMQC to alkaloids prior to 1988. We first split the applications cited into two major categories: terrestrial and marine origin. Perhaps surprisingly, applications in the latter category were more numerous. Next, we attempted to further subgroup applications within the two major categories by one of the fundamental nitrogen-bearing heterocycles that the molecule contains. Here again, we must apologize for the order and subgroupings that we have made if they should offend the sensibilities of any member of the alkaloid chemistry community. The groupings we have made seem logical from our point of view as NMR spectroscopists and furthermore lent themselves to a unified presentation.

9.1 Terrestrial Alkaloids

The earliest application of inverse-detected 2D NMR techniques to a terrestrial alkaloid of which we are aware was the assignment of the ^{1}H and ^{13}C NMR spectra of the indolocarbazole alkaloid staurosporine (**7**) reported by Meksuriyen and Cordell (1988). Later in 1988, Edwards et al. utilized HMQC in the elucidation of the structure of pumiliotoxin-A (**10**) and related indolizine alkaloids isolated from Panamanian poison frogs. These were the only applications of inverse-detected 2D NMR techniques in alkaloid to appear in 1988. We will thus begin with staurosporine (**7**) and proceed to related carbazole-based alkaloids before continuing to pumiliotoxin-A (**10**) and other terrestrial alkaloids.

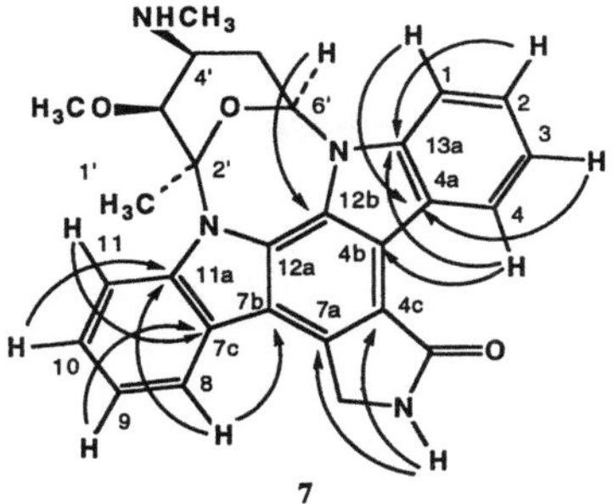

9.1.1 Carbazole Alkaloids

Staurosporine. The indolocarbazole alkaloid staurosporine (**7**) was first isolated from *Streptomyces staurosporeus* Awaya (AM-2282) and subsequently from other actinomycetes. The utilization of inverse-detected techniques with staurosporine was necessitated by the relatively low solubility of the alkaloid in deuterochloroform, which was selected as the solvent in an effort to avoid some of the resonance overlap encountered in DMSO-d$_6$, MeOH-d$_4$, and Me$_2$CO-d$_6$. Although the solubility of **7** in CDCl$_3$ was sufficient for one-dimensional and homonuclear 2D NMR experiments, it was too low for the acquisition of any heteronucleus-detected 2D NMR data. With a 20-mg sample of **7**, Meksuriyen and Cordell (1988) were able to record an HMQC spectrum using only 8 transients/t$_1$ increment using the pulse sequence of Bax and Subramanian (1986). Total acquisition time for the HMQC spectrum was 1.27 h, in contrast to the required 10,000 transients acquired to obtain a ^{13}C reference spectrum, the latter probably requiring overnight data acquisition.

Long-range heteronuclear connectivities were established using the HMBC experiment of Bax and Summers (1986). The authors report optimizing the long-range delay for 7 Hz and the delay for the low-pass J-filter (Kogler et al. 1983) to 139 Hz. We routinely use 140 Hz as a default value for the low-pass J-filter for aralkyl compounds and it works quite well. The 7 Hz optimization of the long-range delay is also quite reasonable. We typically would employ about 6 Hz for this type of molecule in our laboratories.

Long-range connectivities are shown on the structure of staurosporine above and consist exclusively, in this case, of $^3J_{CH}$ correlations. Rather than confusing the presentation, we have shown only correlations to quaternary carbon resonances. In addition to providing the means of assigning quaternary carbon resonances in cases where the assignment of a given proton is known, the connectivities shown also provide the means of orienting the four-spin systems relative to the molecular framework. For example, H4, in a bay-region and deshielded by the carbonyl at the 5-position, would be expected to resonate furthest downfield of the aromatic protons; H4 was assigned at 9.42 ppm. Given this assignment as a starting point, the correlation of H4 to C13a, resonating at 136.6 ppm, orients the four-spin system.

From the COSY spectrum, the sequence of the members of the four-spin system can usually be established. Given the assignment of the H2 resonance from the COSY spectrum, the assignment of C13a is redundantly confirmed by the $^3J_{CH}$ coupling pathway shown.

Jadiffine. Later in 1989, Garnier et al. reported the isolation and structure elucidation of jadiffine **(8)** from *Vinca difformis*. The HMQC experiment

(Bax and **Subramanian** 1986) was used to establish the direct proton-carbon chemical shift correlations in the total assignment of the carbon spectrum of the alkaloid.

Neocarazostatins. The next reported application of inverse-detected 2D NMR to a terrestrial carbazole alkaloid did not appear until 1991 in the report of Kato et al. A series of free radical scavenging compounds, neo-carazostatins A–C, were isolated from a *Streptomyces* sp. strain. The structure of neocarazostatin-A **(9)** is shown below. Long-range connectivities from an HMBC spectrum (not referenced) were used in assigning the attachments of the two side chains to the carbazole nucleus.

9.1.2 Indolizidine Alkaloids

Digressing, we noted above that the other 1988 application of inverse-detected 2D NMR was to indolizine alkaloids of the pumiliotoxin-A and allopumiliotoxin classes (Edwards et al. 1988). A total of ten alkaloids were isolated in small quantities from the skins of poisonous Panamanian frogs. The HMQC experiment was utilized to establish one-bond proton-carbon heteronuclear chemical shift correlations. The structure of one of the new alkaloids described in this report, an 8-methylindolizidine, **10**, is shown.

9.1.3 Diterpene Alkaloids

Another reported application of inverse-detected 2D NMR was the 1989 elucidation of the structure of the diterpene alkaloid lassiocarpine (**11**) (Takayama et al. 1989). This study utilized both HMQC to establish one-bond correlations and HMBC to obtain long-range connectivity information to link the structural components elucidated by other means together.

Connectivities shown on the structure of lassiocarpine (**11**) above allowed the authors to bridge the quaternary carbons and heteroatom in the structure to link structural fragments, allowing the assembly of the hexacyclic ring system.

9.1.4 Indole-Derived Alkaloids

Perhaps the largest single group of terrestrial alkaloids to which inverse-detected 2D NMR techniques have been applied are the diverse indole-derived/containing alkaloids.

9.1.4.1 Heteroyohimbine Alkaloids

Isoreserpiline, Tetraphylline, and Reserpiline Hydrochloride. Inverse-detected 2D NMR was also utilized in 1989 to establish direct-proton carbon shift correlations in three heteroyohimbine alkaloids isolated from *Rauwolfia serpentina* (DeBruyn et al. 1989). The alkaloids discussed in this report included isoreserpiline, tetraphylline, and reserpiline hydrochloride. The structure of one of these alkaloids, isoreserpiline (**12**), is shown.

1 2

9.1.4.2 Indoloquinoline Alkaloids

Three members of the relatively rare indoloquinoline family of alkaloids have been studied by the authors. These include cryptolepine (**13**), quindoline (**3**), and the complex spiro-nonacyclic alkaloid cryptospirolepine (**1**).

1 3

Cryptolepine. The indoloquinoline alkaloid cryptolepine (**13**), isolated from the West African plant *Cryptolepis sanguinolenta*, is of interest for its use as an endemic medicinal for the treatment of malaria in West Africa. The proton and carbon NMR spectra of this alkaloid were first assigned in $CDCl_3$ using heteronucleus-detected 2D NMR techniques by Ablordeppy et al. (1990). We subsequently utilized HMQC and HMBC to assign the spectra of **13** in DMSO-d_6, with the intent of having assignments in a solvent capable of dissolving a wider range of analogs which were then being isolated (Tackie et al. 1991). Connectivities to the quaternary carbon resonances from the HMBC spectrum are shown on the structure.

Our study of cryptolepine, performed on 5 mg of material dissolved in 0.8 ml of DMSO-d_6, highlights what will probably become an increasingly common protocol as inverse-detected 2D NMR methods become more widely implemented. Typically, we acquire a proton reference spectrum followed by COSY and HMQC spectra. On the sample in question here, usable data for these three experiments can be acquired in under an hour. We followed the HMQC spectrum with the overnight acquisition of an HMBC spectrum optimized for 10 Hz (50 ms), preferring to acquire the considerable connectivity information for the HMBC spectrum prior to the acquisition of a carbon reference spectrum. This procedure, of course, necessitates a guess on the part of the spectroscopist of the range of quaternary carbon chemical shifts in the molecule being studied. In this study we "guesstimated" a chemical shift range of 110–160 ppm as sufficient to acquire the spectrum, ignoring, of course, the N-methyl resonance upfield. We were almost correct in our parameterization; only the C9a resonance at 159.95 ppm caused problems, because it was located at the very edge of the F_1 spectral window. Alternatives available at this point include: reacquiring the HMBC spectrum, investing a second night of instrument time, or using the SIMBA experiment (Crouch and Martin 1991) – the selective one-dimensional inverse-detected analog of the HMBC experiment described above. The latter is clearly the more time-efficient method, since it confirmed connectivity information for the C9a resonance in about 5 min. Rationales for the utilization of SIMBA spectra are presented above in the experimental techniques section.

Practically, SIMBA spectra of C5a and C4a confirmed the correlation from the N-methyl resonance that was excluded from the F_2 spectral window, in about 10 min. A series of SIMBA spectra using a range of five optima were also acquired for the C10a resonance did not exhibit a $^2J_{CH}$ correlation to the H11 resonance of cryptolepine. Unfortunately, the SIMBA spectra likewise failed to confirm this correlation pathway, but allowed the complete study to be undertaken in less than one half hour. Acquisition of the corresponding set of HMBC spectra to look for the $^2J_{CH}$ connectivity from H11→C10a would have been time prohibitive.

Quindoline. A second known alkaloid, quindoline (**3**), was also isolated from *C. sanguinolenta*. No assignment data existed for quindoline other than the low-field data contained in the initial report of the isolation of the alkaloid (Dwuma-Badu et al. 1978). Unequivocal proton and carbon resonance assignments were derived from HMQC and HMBC spectra. Long-range connectivities observed in the HMBC spectrum of quindoline are shown on the structure.

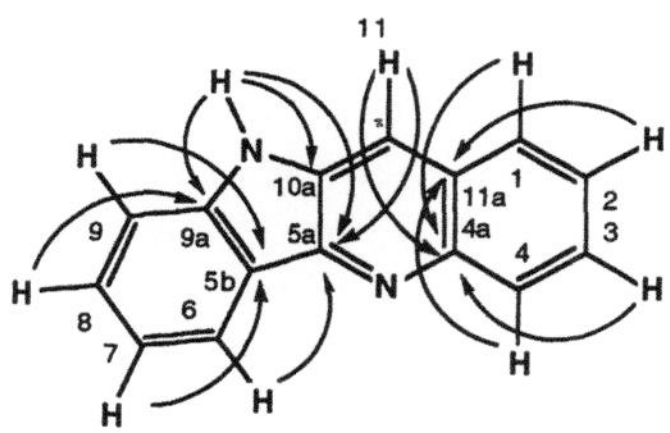

3

Cryptospirolepine. The elucidation of the structure of the novel, spiro-nonacyclic alkaloid cryptospirolepine (**1**) made extensive use of inverse-

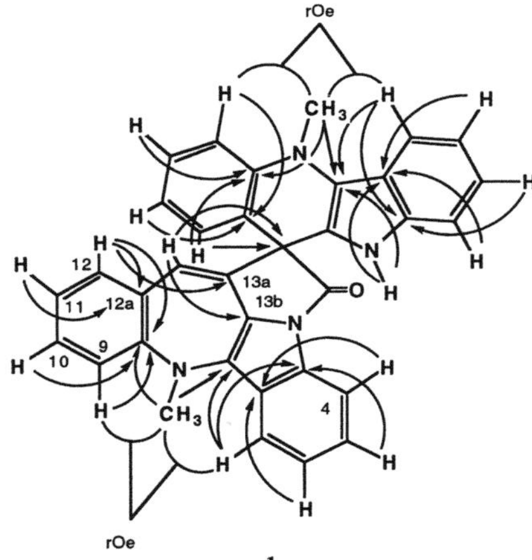

detected methods (Tackie et al. 1993). Direct proton-carbon chemical shift correlations were established from an HMQC spectrum. Substructures of the alkaloid were assembled using HMQC-TOCSY to identify the component resonances of the four-spin systems (see discussion above), HMBC to link the four-spin systems to the quaternary carbon resonances and ROESY data to link individual spin systems to the N-methyl groups. Connectivities to quaternary carbon resonances in the HMBC spectrum are shown on the structure.

Although most of the connectivities are via $^3J_{CH}$, there were several $^2J_{CH}$ and one $^4J_{CH}$ correlation. The former are easily identified and simply provide an additional level of redundancy in the assignment of the quaternary carbon to which they correlate. An example of one of these correlations was the coupling of H12 to the C12a resonance. This coupling is in addition to the anticipated correlations from H9 and H11 to C12a. More interesting, and also more frustrating during the elucidation of the structure of cryptospirolepine, was the strong $^4J_{CH}$ correlation from H12 to C13a at the juncture of the N-methylazepine and γ-lactam rings. We made repeated efforts to interpret this correlation as a normal three-bond coupling, proposing numerous erroneous structural possibilities that were discounted one by one.

The elucidation of the structure of cryptospirolepine (**1**) also made use of $^1H-^{15}N$ HMQC data and, as such, is one of the few examples of the use of this experiment in alkaloid chemistry to date (see also Sect. 9.2.8 below for a second example). The proton resonating at 10.67 ppm in the spectrum of **1** could arise from either an −OH or a −NH−. The origin of this resonance as the latter was confirmed by a one-dimensional $^1H-^{15}N$ HMQC spectrum in a few minutes. A full 2D $^1H-^{15}N$ spectrum was recorded, the ^{15}N

resonating at 118.9 ppm. The chemical shift of the protonated ^{15}N resonance of **1** compares favorably with several indoloquinolizidine alkaloids and reserpine, which have ^{15}N resonances in the range from 117.9 to 119.6 ppm (Fanso-Free et al. 1979). While the use of ^{1}H–^{15}N HMQC spectra has, thus far, been quite limited, additional examples of the use of this technique will undoubtedly appear. The sensitivity advantage of the ^{1}H–^{15}N HMQC experiment relative to direct observation of ^{15}N is nearly a factor of 10^{3}, which makes it feasible to acquire ^{1}H–^{15}N spectra in reasonable periods of time. For the reader interested in ^{15}N chemical shifts of alkaloids, some data are to be found in the monograph of Levy and Lichter (1979) although this is rather dated. More recent reviews on the topic have also appeared (von Phillipsborn and Müller 1986; Witanowski et al. 1993) and may be of more use to individuals interested in performing these experiments. In our opinion, however, the only practical way to access ^{15}N chemical shift information is through the acquisition of inverse-detected ^{1}H–^{15}N HMQC spectra.

Finally, cryptospirolepine also provided the first "real-world" application of the SIMBA experiment in the elucidation of the structure of a new alkaloid. A SIMBA experiment was acquired in which the C13b carbon resonating at 120.14 ppm was selectively pulsed, allowing the observation of a weak $^{4}J_{CH}$ coupling to the H4 resonance and confirming the presence of an indolobenzazepine component to the structure. The particulars of this experiment have been discussed above in the experimental section of this chapter and the spectrum is shown in Fig. 16.

9.1.4.3 Strychnos Alkaloids

Holstiine. Although the strychnos alkaloids are a large and diverse group, only one application of inverse-detected 2D NMR methods to a member of this alkaloid family has been reported. Holstiine (**14**) was originally reported by Bosly (1951) to contain a dihydroxypiperidinone structural moiety. Subsequently Bisset et al. (1975) revised the structure to the hydroxyoxazepinone containing structure shown on the basis of low-field NMR data. Cherif et al. (1990) finally confirmed the structure of holstiine as that shown by **14**

14

through the use of HMQC and HMBC spectra. Unequivocal assignments for all resonances were reported; stereochemical differentiation of geminal methylene resonances were based on responses in a NOESY spectrum. It is interesting that although HMBC spectra were acquired at 50, 83, and 125 ms (corresponding to 10, 6, and 4 Hz, respectively) the spectra were all largely comparable with the 6 Hz spectrum showing only a few additional responses relative to the 10 Hz spectrum. It should also be noted that the quality of the 4 Hz data was inferior to the other spectra, probably due to losses of signal during the significantly longer delay required for the experiment.

9.1.4.4 Oxindole Alkaloids

Macroxine. The sole application of inverse-detected 2D NMR techniques to an oxindole was in the elucidation of the structure of macroxine (**15**)

1 5

reported by Atta-ur-Rahman et al. (1991). Connectivities from H5, H6, H3, and H20 were critical to the determination of the structure. Correlations from H6 a/b to C5, C7, C8, and C2 and from H3a to C8, for example, confirmed that both C3 and C6 were linked to the 7-position. Other structural components were linked together in a similar fashion using long-range connectivities contained in the HMBC spectrum.

9.1.4.5 Aspidospermine Alkaloids

1,2-Dehydroaspidospermine. An interesting application of inverse-detected 2D NMR techniques was reported in conjunction with a study of the flow thermolysis of 1,2-dehydroaspidospermidine (**16**) by Hugel et al. (1991).

1 6

The authors isolated four thermolysis products in small quantities, the structures of which were elucidated using a combination of HMQC and HMBC spectra. Three of the four products of the study, **17**, **18**, and **20**, are novel alkaloidal systems not yet reported in nature; **17**, **18**, and **19** from two successive [1,5] sigmatropic shifts, the fourth compound, **20**, surprisingly, can be formed only after four successive [1,5] sigmatropic shifts. Extensive use of HMBC connectivities was made in establishing the structures of **17–20**. The reader interested in this class of alkaloids will find the work of Hugel et al. (1991) a useful starting point for spectral assignment and/or structure elucidation studies.

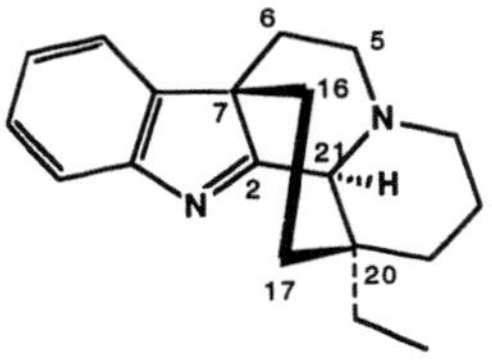

1 7

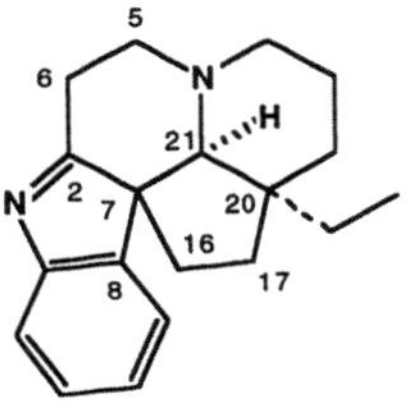

18

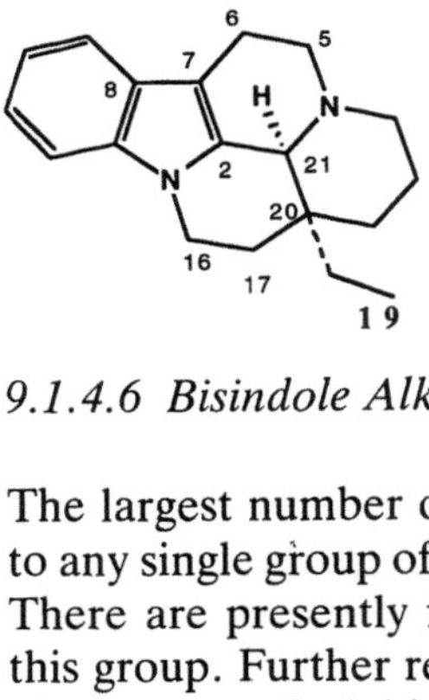

1 9

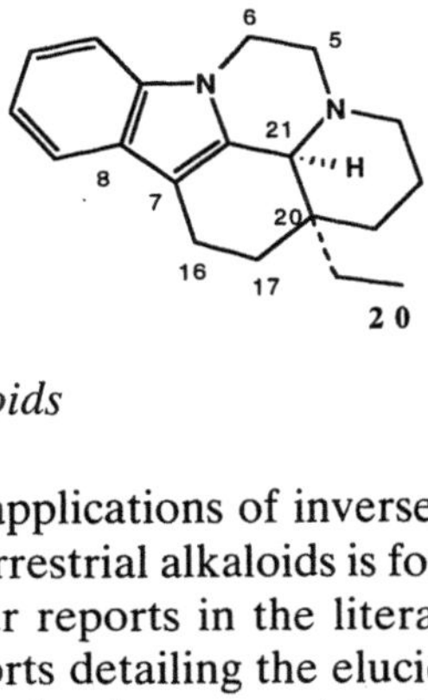

2 0

9.1.4.6 Bisindole Alkaloids

The largest number of applications of inverse-detected 2D NMR techniques to any single group of terrestrial alkaloids is found for the bis-indole alkaloids. There are presently four reports in the literature pertaining to members of this group. Further reports detailing the elucidation of structures of this class of complex alkaloids or in the spectral assignment of known members of the class will undoubtedly continue to appear. These molecules frequently exhibit highly congested spectra in which HMQC-TOCSY may be advantageous. Correlations from the HMBC spectrum are also frequently capable of linking the two structural constituents of a bis-indole to one another.

Pericyclivine Analogs. Massiot et al. (1990) have utilized both HMQC and HMBC spectra in assigning the structures of products of synthetic coupling reactions designed to afford bisindoles – undulatine specifically. Treatment of a pericyclivine derivative afforded an intermediate that could be coupled with cabucraline to afford **21**. The spectra of **21** were totally assigned from the inverse-detected spectra. No details of the optimization are given and only a few of the connectivities observed in the HMBC spectrum were

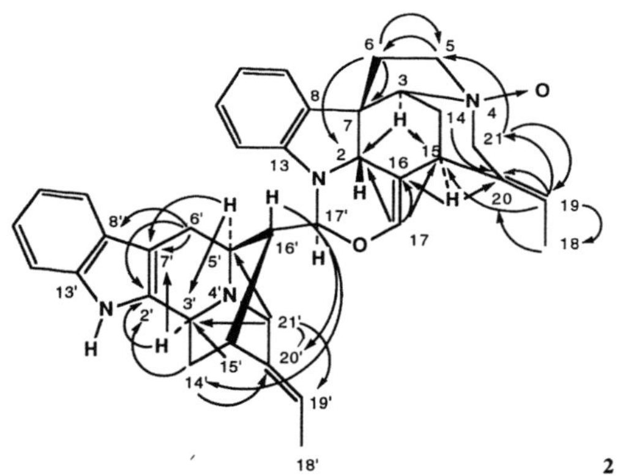

2 1

considered noteworthy by the authors; these are shown on the structure and, usefully, confirm the linkage between the two molecular subunits.

Divaricine. Mukherjee et al. (1991) reported the elucidation of the structure of divaricine (**22**), a new bisindole from *Strychnos divaricans*. Again, HMQC

2 2

was employed to establish direct proton-carbon shift correlations and extensive utilization was made of HMBC in confirming points of attachment of various structural components that could be elucidated from other experiments such as COSY and homonuclear TOCSY spectra. As an example of the use of HMBC, the C3–C14 bond in the indoline portion of the molecule was indirectly established through a series of observed connectivities. Correlations of C2 to H3, the H6 methylene protons, and H17 and correlation of C15 to H3, H17, H19, and one of the 21 methylene protons established, albeit indirectly, the requirement of a C3–C14 bond for the structure to be consistent with the observed connectivities. In similar fashion, other correlations in the indoline portion of the structure were established.

In contrast to the establishment of the indoline portion of the molecule from HMBC connectivities, the balance of the carbon resonances suggested the presence of a sarpagine-like framework. Chemical shift comparison of divaricine to normacusine-B O-acetate showed the two molecules to have nearly identical ^{13}C chemical shifts. Subsequent utilization of the connectivities observed in the HMBC spectrum confirmed the suspected structural similarity, completing the elucidation of the structure. It is interesting, however, that neither H16′ nor H17′ was reported to exhibit any correlation between the two halves of the molecule.

Crooksiine. Another bis-indole alkaloid studied using inverse-detected techniques in 1991 was crooksiine (**23**) (Alam and Mroue 1991) isolated

from *Haplophyton crooksii* collected in Texas. The authors report using HETCOR to establish the direct ^{1}H–^{13}C shift correlations and an HMBC spectrum to establish the long-range correlations. The latter experiment was performed by Alam and Mroue using an older spectrometer modified by one of the authors of this chapter to perform the HMBC experiment but incapable of broadband heteronucleus decoupling to facilitate the establishment of direct correlations through an HMQC experiment. Although Alam and Mroue report a total assignment of the carbon NMR spectrum of the alkaloid through the information contained in the HMBC spectrum, only partial connectivity information was provided in the report. The connectivities reported are shown on the structure.

Navelbine^{TM}. The most recent application of inverse-detected 2D NMR methods to a bis-indole was the reported assignment of the proton and carbon NMR spectra of Navelbine^{TM} (**24**) reported by the authors (Spitzer et al. 1992). Direct proton-carbon shift correlations were established using an HMQC spectrum. The authors also reported the comparative evaluation of the recently reported GEM-COSY technique (Domke et al. 1991), which affords a heteronuclear correlation spectrum of only the methylenes of a molecule, with the HMQC spectrum. Based on their assessment, GEM-COSY, while useful, did not offer any advantage over the HMQC experiment. In addition, the nature of the experiment required high levels of

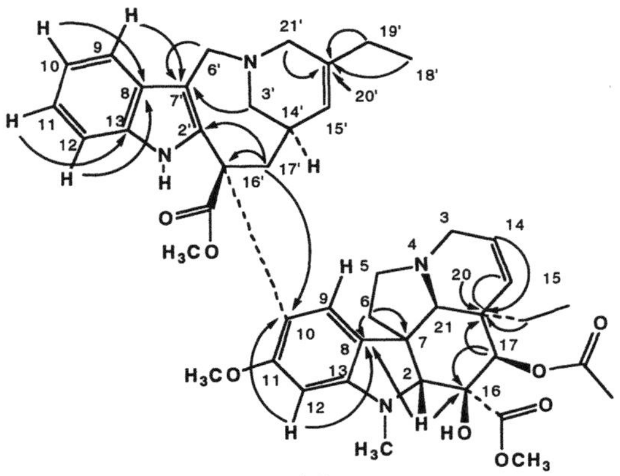

24

digitization, which may serve as an impediment to some investigators. Furthermore, although no direct comparison between GEM-COSY and DEPT-HMQC (Kessler et al. 1989b) has been reported, it is probable, in the opinion of the authors, that the latter will prove to be superior. Long-range connectivities were obtained from an HMBC spectrum and are shown on the structure. A single long-range correlation was observed between the two halves of the alkaloid; H17' in the velbanamine segment was long-range correlated with the C10 resonance in the vindoline portion. No vestige of a correlation was observed between H9 and C16'.

Perhaps, from the standpoint of applications, the most interesting facet of the reported spectral assignment of Navelbine™ was the utilization of HMQC-TOCSY to circumvent the overlap of two vinyl protons, H14 and H15', both of which resonate at 5.76 ppm. From the HMQC spectrum, the resonances at 5.76 ppm correlated with carbons resonating at 124.0 and 122.5 ppm. An HMQC-TOCSY spectrum with a 14-ms mixing period showed the resonant pair at 5.76/124.0 ppm to relay to another vinyl proton resonating at 5.24 ppm and to one member of an anisochronous methylene pair resonating at 3.21 ppm. The mixing time in this case is too short for a double transfer to occur and hence, the protons to which magnetization was relayed must flank the proton resonance in question. On this basis, the 5.76/124.0 ppm pair is assigned as H14/C14; the vinyl and methylene proton are assigned as H15 and H3, respectively. The remaining vinyl proton-carbon pair resonating at 5.76/122.5 ppm may thus be assigned as H15'/C15'.

9.1.5 Pyrrole, Imidazole, Pyrrolizidine, Pyridine, and Related Alkaloids

The alkaloids remaining to be described at this point have been grouped together for convenience although they are, to our minds as spectroscopists, unrelated.

Oxalicine A. An interesting pyridine-containing alkaloid, oxalicine-A (**25**), was communicated in 1989 (Ubillas et al. 1989). The communication referred to the HMQC experiment but cited the HMBC reference of Bax and Summers (1986). Unfortunately, no details of any long-range connectivities were communicated and, to the best of our knowledge, have not been reported elsewhere.

2 5

Mikimopine. The alkaloid mikimopine (**26**), induced by *Agrobacterium rhizogenes*, was isolated from the hairy roots of tobacco (Isogai et al. 1990). On treatment with acid, the alkaloid underwent dehydration to afford the corresponding γ-lactam, which was characterized using connectivities from an HMBC spectrum. Connectivities observed are shown on the structure.

26

Curassanecine. In a synthetic study, Gramain et al. (1991) report the preparation of both natural curassencine (**27**, stereochemistry shown) and its 1-epimer. This communication is interesting because both $^2J_{CH}$ couplings were observed in the naturally occurring alkaloid from the H2 methylene protons to C1, but only a correlation from the upfield H2 methylene proton to C1 was observed in the case of the 1-epimer. Ultimately, the stereochemistry

2 7

was established by the acquisition of an X-ray crystal structure of a 1-phenyl analog. Interestingly, the same behavior in terms of $^2J_{CH}$ couplings from H2→C1 was observed for the 1-phenyl analogs, and it was on the basis of this long-range coupling behavior that the stereochemistry of **27** was assigned.

Lydicamycin. The final application of inverse-detected 2D NMR techniques to a terrestrial compound was to lydicamycin (**28**), a compound reported

2 8

(Hayakawa et al. 1991a,b) as an antibiotic, although it has alkaloidal attributes. We chose to include this compound here to illustrate an interesting utilization of HMBC data to establish the structure of a long, isoprene-derived chain connecting two portions of the molecular structure.

A number of structural fragments of lydicamycin were elucidated using conventional homonuclear COSY and TOCSY spectra. The isoprene-derived partial structures were linked together using long-range connectivities from the methyl groups in an HMBC spectrum. Connectivities from methyl groups are among the easiest of any long-range correlations to observe in an HMBC spectrum. Thus, as illustrated by the connectivities shown on the structure, the methyl groups, several located at quaternary vinyl carbons, were used to couple structural fragments together, linking components from one isoprene component to the next with the quaternary carbon resonance serving as the linkage-point. Other connectivities observed in the HMBC spectrum are also shown on the structure but do not warrant further comment.

9.2 Marine Alkaloids

It is interesting that applications of inverse-detected 2D NMR experiments in the elucidation of marine alkaloid structures presently outnumber terrestrial applications. Most probably, this situation has arisen because there is only a relatively short history and limited body of knowledge associated with alkaloids of marine origin. While terrestrial alkaloids fall into families that have been studied for many years, marine alkaloids, in contrast, have the potential to be without precedent each time a new compound is isolated. Consequently, in our opinion, it is probable that investigators studying marine alkaloids have had to resort to more powerful NMR techniques sooner than their colleagues dealing with terrestrial alkaloids.

Although probably not significant, it is interesting that the first applications of inverse-detected 2D NMR techniques in the study of marine alkaloids were applied to quinoline-based alkaloids. There have also been a sizeable number of isoquinoline and acridine-derived marine alkaloids reported. Thus, we will begin our survey of marine alkaloids with these groups.

9.2.1 Quinoline Alkaloids

Cystodytin A–D. Tunicates have been a rich source of marine alkaloids. The first reported application of inverse-detection to an alkaloid of marine origin of which we are aware was the elucidation of the structures of cystodytin A–D from the Okinawan tunicate *Cystodytes dellechiajei* reported by Kobayashi et al. (1988). The structure of cystodytin-A (**29**) is shown. The

2 9

authors made extensive use of long-range connectivities from an HMBC spectrum in the elucidation of the nucleus of this alkaloid.

Structurally, the cystodytins differ only in the alkyl substituent attached to the side-chain nitrogen at the 14-position. Connectivity information from COSY and conventional HETCOR spectra can easily be seen to establish a four-spin, a two-spin, and an isolated aromatic/vinyl proton resonance. Using the four-spin system as a starting point, connectivities from H2 and H4 to a quaternary carbon resonating at 145.0 ppm allow the inference of a nitrogen. By utilizing connectivities from H4 to C4b and H5 to C4a, the relationship of the four-spin system to the two-spin system is established. Furthermore, based on the chemical shift of the H6 resonance, the location of a second nitrogen atom is also fixed. Correlations from H6 and H9 to C7a locate the isolated aromatic resonance relative to the two-spin system. In this systematic fashion, the authors assembled the nucleus of the alkaloid that is comprised of two fused quinoline-derived systems. Correlations from H9–C12 and H12–C10a also serve to attach the side chain the nucleus of the alkaloid.

Prianosins B–D. A group of novel sulfur-containing alkaloids were isolated from the Okinawan marine sponge *Prianos melanos.* The first member of this family of alkaloids was communicated in 1987 by Kobayashi et al. (1987), the structure confirmed by X-ray. Structures of prianosins B–D were established, following the precedent of prianosin A, by 2D NMR methods. A combination of homonuclear COSY and TOCSY spectra were used to establish the proton spin systems; direct proton-carbon shift correlations were established using a conventional HETCOR spectrum. The extensive use of long-range connectivities from an HMBC spectrum by Cheng et al. (1988) is shown on the structure of prianosin C (**30**). Cor-

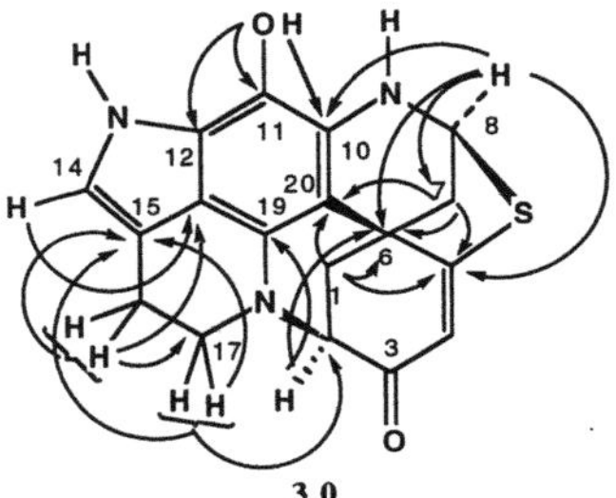

relations associated with the protonated positions flanking the spiro-center were noted by the authors to be particularly useful in establishing correlations in this region of the skeleton of the molecule.

Wakayin. The next marine quinoline-derived alkaloid to be reported was wakayin (**31**), the first pyrroloiminoquinone isolated from an ascidian, *Clavelina* sp. (Copp et al. 1991). Direct proton-carbon shift correlation was established using the HMQC experiment. Long-range connectivities were established using both HMBC and INAPT (Bax et al. 1985b). Relatively little discussion of the long-range connectivity information was presented in the report. One key connectivity mentioned was the H13–C16 correlation shown, establishing the point of attachment of the 3-indolyl substituent. The complementary correlation from H17–C14 was not observed; the authors suggest that this may have been due to the broadening of C14 ($v_{1/2}$ = 9.9 Hz), which would lead to relatively low sensitivity for this carbon. It is interesting to speculate that the SIMBA experiment might have been used to advantage in this case, since high levels of signal-to-noise could be achieved with this one-dimensional inverse-detected experiment.

9.2.2 Acridine Alkaloids

Three reports in 1989 and 1990 described the structures of three closely related acridine-derived alkaloids. Isolations were from a tunicate, a sponge, and, interestingly, a tunicate and its prosobranch mollusc predator.

Shermilamine A–B. The first of the acridine-derived alkaloids to be reported were shermilamine A and B (**32**) (Carroll et al. 1989). Using long-range

connectivities from the HMBC spectrum, it was possible to completely assemble the carbon skeleton. Correlations to quaternary carbons are shown on the structure. The elucidation of the structure was analogous to the procedure used by Kobayashi et al. (1988) in elucidating the structures of the cystodytins (**29**) described above.

Nordercitin, Dercitamine, Dercitamide, and Cyclodercitin. A series of three pyridoacridine alkaloids related to the shermilamines were reported by Gunawardana et al. (1989). In assigning the structures of nordercitin (R = NMe$_2$) dercitamine (R = NHMe), and dercitamide (R = NHCOEt) (**33**), isolated from two deep-water marine sponges of the family Pachastrellidae, the authors used a combination of COLOC and HMBC spectra. The report did not compare results of the two experiments nor unfortunately did it

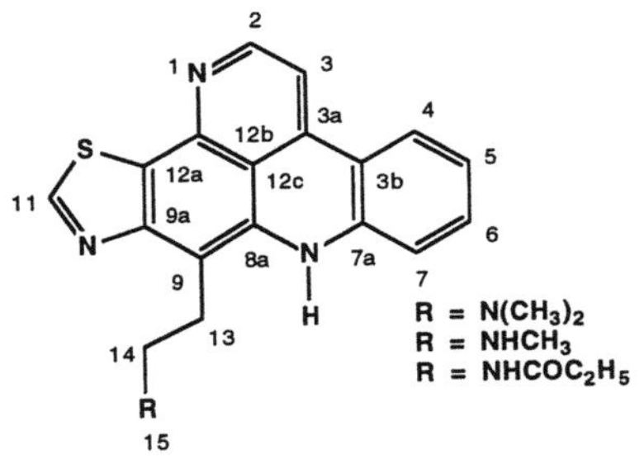

3 3

R = N(CH$_3$)$_2$
R = NHCH$_3$
R = NHCOC$_2$H$_5$

differentiate which correlations were observed in a particular experiment. Connectivities were observed to all carbons except C12c, which is surprising in that C12c should have been correlated to the H3 resonance, as in the case of the shermilamines (**32**).

A fourth pyridoacridine alkaloid, cyclodercitin, was also described in the report. The alkaloid was poorly soluble in methanol-d$_4$ and correlations could not be observed from the methylene resonances in an HMBC spectrum. Cyclodercitin did, however, readily dissolve in TFA-d, albeit with conversion of the pyrrolidine to afford the corresponding pyrrole, **34**. Correlations were observed from the H9 resonance of the pyrrole to C7a and C13d, confirming the site of fusion of the pyrrole to the pyridoacridine nucleus.

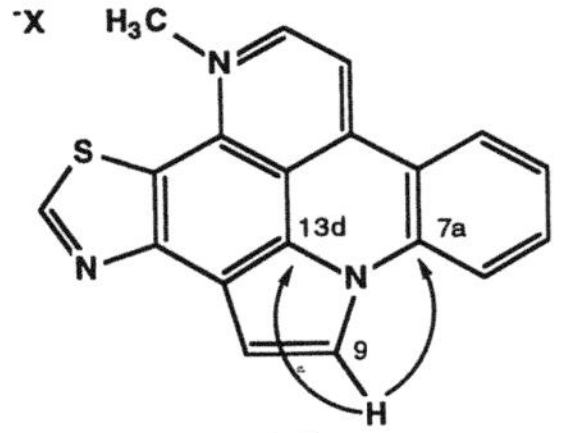

3 4

Kuanoniamines A–D. The third group of acridine-derived alkaloids to be reported, was the kuanoniamines (**35, 36**) (Carroll and Scheuer 1990) isolated from a Micronesian tunicate and its prosobranch mollusc predator *Chelynotus semperi*. Structurally, the kuanoniamines differ from nordercitin and the related compounds discussed above only in the fusion of the thiazole to the pyridoacridine nucleus. Direct proton-carbon shift correlations were

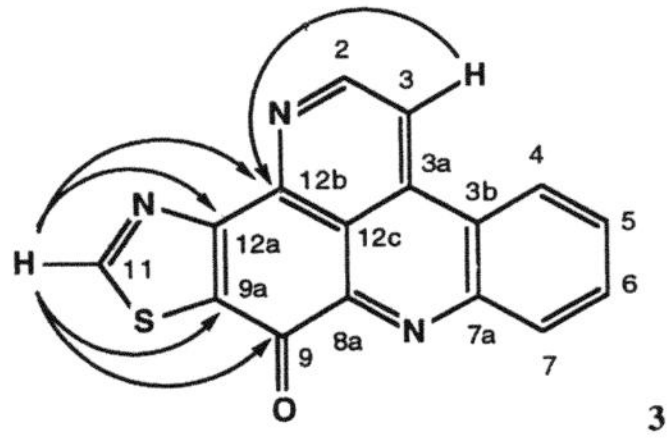

3 5

3 6

established using the HMQC experiment. Long-range connectivities were obtained from a 5-Hz optimized HMBC spectrum. Connectivity pathways for the quaternary carbons were essentially the same as those reported for the shermilamines (**32**). A number of atypical $^4J_{CH}$ correlations were, however, reported and are shown on the structure of kuanoniamine A (**35**). Correlations fixed the side-chain attachment at the 9-position of the pyridoacridine system and were typical $^3J_{CH}$ couplings.

9.2.3 Tetrahydroisoquinoline Alkaloids

Ecteinascidins. Although only two reported applications of inverse-detected 2D NMR methods to isoquinoline alkaloids appear in the literature, these represent perhaps the most interesting and complex marine alkaloids reported to date. In back-to-back papers, Wright et al. (1990) and Rinehart et al. (1990) reported the structures of a series of ecteinascidins. The two research groups reported the elucidation of analogs of the same sulfur-containing tristetrahydroisoquinoline alkaloids. The authors utilized HMQC to establish direct proton-carbon shift correlations and attempted to use the standard HETCOR experiment optimized for 6–10 Hz to observe long-range couplings which can be problemsome (Martin and Zektzer 1988). This effort was unsuccessful and prompted the use of HMBC data. Structural fragments were identified using homonuclear 2D experiments, which were subsequently linked together using connectivity information from the HMBC spectrum. Numerous long-range correlations were observed in the HMBC spectrum and are shown on the structure of ecteinascidin 743 (**37**), one of the compounds reported in the work of Rinehart et al. (1990).

9.2.4 Pyrrole and Imidazole-Derived Alkaloids

A number of pyrrole- and imidazole-containing marine alkaloids have had inverse-detected 2D NMR experiments applied in the elucidation of their

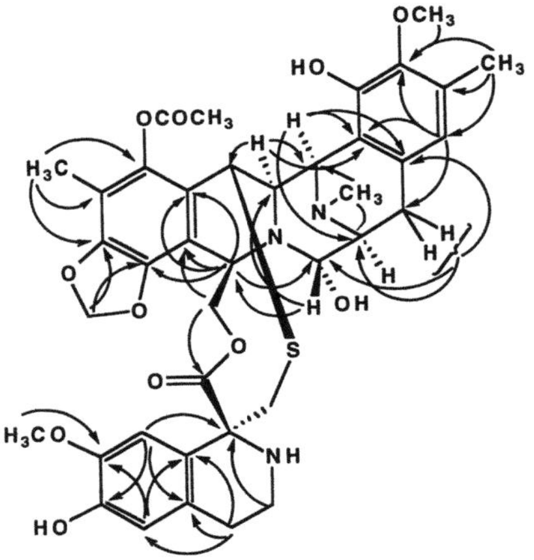

3 7

structures. Unlike the acridine-derived alkaloids discussed above, none of these compounds are structurally related.

Naamidine-D. A number of 2-amino imidazole alkaloids were reported from the marine sponge *Leucetta chagosensis* (Carmely et al. 1989). An HMQC spectrum was utilized to establish direct proton-carbon shift correlations for naamidine-D (**38**), which was isolated in minute amount.

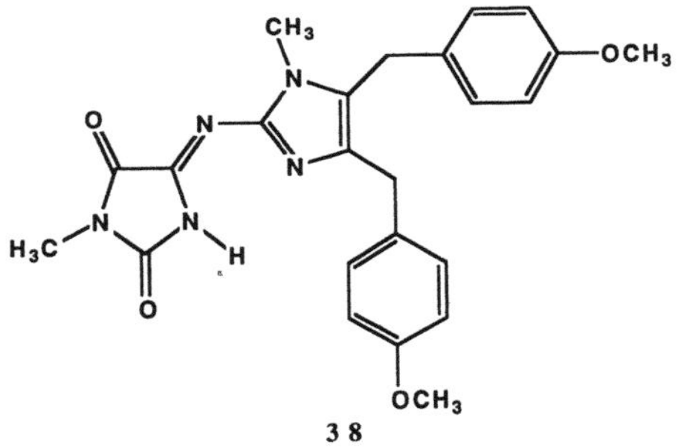

3 8

Although the authors refer to establishing direct correlations, the Bax and Summers (1986) HMBC paper was referenced.

1,2,3-Trithiane. A very novel imidazole-derived alkaloid, **39**, containing an unusual 1,2,3-trithiane ring was also reported in 1989 (Copp et al. 1989) isolated from a New Zealand ascidian, *Aplidium* sp. This group used a conventional HETCOR experiment to establish one-bond proton-carbon shift correlations followed by an HMBC experiment optimized for 8.3 Hz to obtain the long-range correlations. Long-range correlations were used to link the phenyl and imidazole substituents to the quaternary carbon of the 1,2,3-trithiane ring.

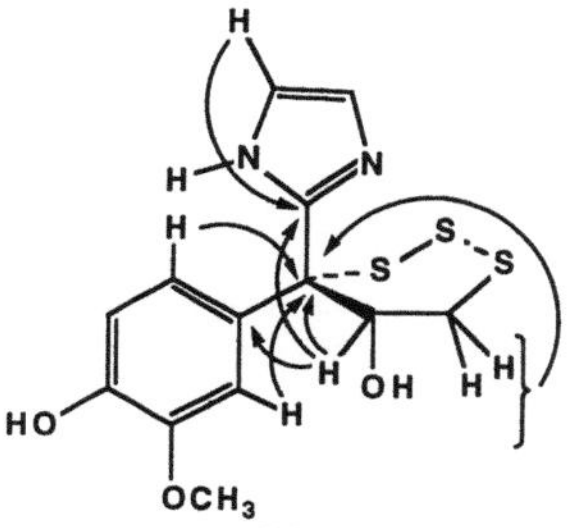

Discodermide. The next report of a pyrrole/imidazole-derived alkaloid to appear was discodermide (**40**), a complex macrocyclic lactam isolated from the marine sponge *Discodermia dissoluta* (Gunasekera et al. 1991). Structural fragments containing contiguous protonated carbons were first established using COSY, LR-COSY (using fixed delays to select for long-range proton-proton couplings) and RCOSY. Long-range connectivities from an HMBC spectrum were then utilized to link the established structural fragments together through quaternary carbons or across heteroatoms. Some of the long-range connectivities reported in this paper are shown on the structure. One other point worth noting was the authors' usage of one-bond heteronuclear coupling constants for the C3 and C4 resonances of the oxirane. Couplings of $^1J_{C3H} = 180$ and $^1J_{C4H} = 182\,\text{Hz}$ were used to infer the presence of the epoxide in the structure. Readers should recall that easy access to one-bond heteronuclear couplings can be had through the proton spectrum by simply performing an HMQC experiment one-dimensionally without decoupling. In this way, proton resonances appear as doublets split by the one-bond heteronuclear coupling constant. Even in highly congested regions where it would be necessary to perform a ^{13}C-coupled HMQC experiment, this approach can provide the one-bond heteronuclear coupling constants far more time-efficiently than the direct acquisition of a proton-coupled ^{13}C spectrum.

Manzacidins A–C. A trio of novel pyrrole-derived alkaloids containing a tetrahydropyrimidine in their structures, manzacidins A–C, (**41**) were also

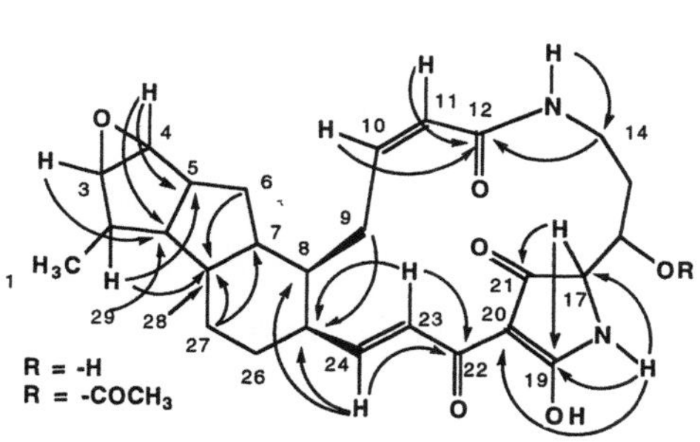

4 1

reported by Kobayashi et al. (1991b), isolated from the Okinawan marine sponge, *Hymeniacidon* sp. The authors used an HMQC experiment to establish one-bond correlations followed by an HMBC experiment to link protonated structural fragments through quaternary carbons and across heteroatoms. Long-range connectivities observed are shown on the structure.

9.2.5 Pyridine-Containing Alkaloids

Several early reports of pyridine-containing alkaloids have appeared describing the application of inverse-detected 2D NMR methods. In these cases, HMBC connectivities were utilized principally to link the pyridine ring of the alkaloid to long alkyl chains.

Haliclamines A, B. Fusetani et al. (1989) reported the isolation and elucidation of the structures of a pair of tetrahydropyridine-containing macrocyclic alkaloids, haliclamines A and B (**42**), from a sponge of the

4 2

genus *Haliclona*. Using COSY and HETCOR spectra, the authors first established the structures of three structural components (A–C on the structure). Long-range connectivities from the HMBC spectrum were then utilized to link the components together, establishing the macrocyclic structure.

Niphatesines A–D and Theonelladins A–D. A group of pyridine alkaloids, **43**, isolated from the Okinawan marine sponge, *Niphates* sp. have been reported by Kobayashi et al. (1990c). An HMQC spectrum was utilized to establish the direct proton-carbon shift correlations followed by an HMBC

4 3

spectrum. The latter provided connectivities which allowed the linkage of the alkyne-containing side chain to the 3-position of the pyridine nucleus. Related pyridine alkaloids, theonelladins A–D, were reported a year earlier by Kobayashi et al. (1989). These compounds differ from the niphatesines in the constitution of the aliphatic side chains. Two contained an alkene function at the 9-position; two were fully reduced and a carbon shorter.

9.2.6 Indole Alkaloids

In contrast to numerous reports of terrestrial alkaloids, only a single report applying inverse-detected 2D NMR methods to indole alkaloids from a marine source has appeared. Kobayashi et al. (1990b) reported the structures of several simple indole alkaloids, hyrtiosins A and B (**44**, **45**), from

4 4   **45**

the Okinawan marine sponge *Hyrtios erecta*. AN HMQC spectrum was used to obtain the direct proton-carbon shift correlations, and an HMBC spectrum provided the long-range correlations. Connectivities to the quaternary carbons are shown on the structures.

9.2.7 Carbazole Alkaloids

Eudistomins B–D. A group of three carbazole alkaloids were isolated from the Okinawan marine tunicate, *Eudistoma glaucus*, and their structures determined using inverse-detected methods (Kobayashi et al. 1990a). The structure of eudistomin B (**46**) is shown. Correlations to the quaternary carbons were used to locate unequivocally the bromo-substituent at the 6-position and to attach the phenethylamino side chain at the 1-position.

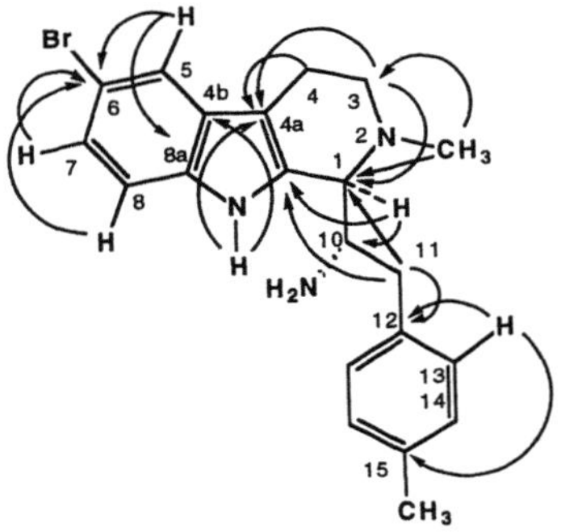

4 6

9.2.8 Oxazole/Thiazole Containing Alkaloids

Tantazoles. Several unusual cytotoxic alkaloids were isolated from the blue-green alga, *Scytonema mirabile* (Carmeli et al. 1990). A combination of HMQC and HMBC data were utilized in the elucidation of the structure, but no details of the long-range connectivities were given in the paper. The structure of didehydrotantazole-A (**47**) is shown. The authors comment

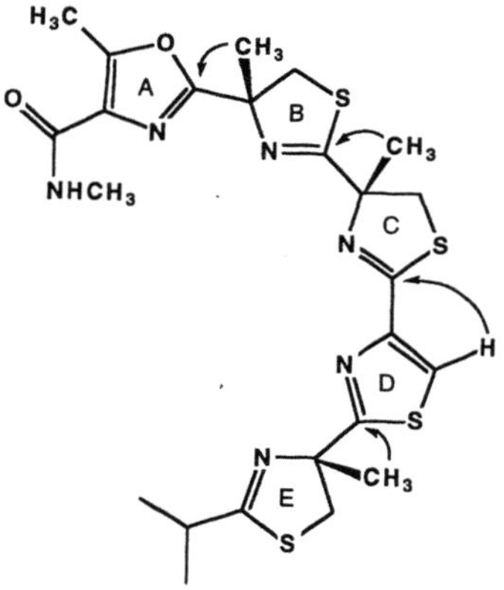

4 7

that the sequence of the rings was established from the HMBC data. Connectivities from the methyl substituents on rings B, C, and E will readily afford two structural fragments consisting of A–B–C and D–E rings. Using the vinyl proton on ring D of the D–E fragment, the linkage of ring C to D can be established. These connectivities are shown on the structure, but these long-range coupling pathways have been inferred by us, not cited by the authors of the paper.

The authors were unable to elucidate the structure of ring A from the HMQC and HMBC data available. To determine the structure of ring A, the authors grew the alga in controlled media affording material that was 82% ^{13}C and >90% ^{15}N enriched. Using a 5-mg sample of the labeled material, the authors were able to determine the structure of ring A from an INADEQUATE (Bax et al. 1980) spectrum. Other than our own work on the structure of cryptospirolepine (**1**) described above, this is the only other paper to our knowledge that details the use of inverse-detected ^{1}H–^{15}N correlation experiments. An HMBC experiment was used since none of the nitrogens in the molecule are protonated.

9.2.9 Bromotyrosine/Bromodopamine-Derived Alkaloids

Purealidin B, C. The structures of several bromotyrosine-derived alkaloids were reported from the Okinawan marine sponge, *Psammaplysilla purea* by Kobayashi et al. (1991c). Structurally, purealidin B (**48**) is perhaps the most

4 8

interesting. The authors report using only HMQC data in the elucidation of this structure.

Long-range connectivities from an HMBC spectrum were utilized to establish the structure of the side-chain of purealidin C (**49**). These connectivities are shown on the structure and illustrate the ability of long-range

4 9

connectivities to bridge several contiguous heteroatom/quaternary carbon centers.

Aplaminone and Neoaplaminone. Three brominated, cytotoxic alkaloids have been isolated from the marine mollusc, *Aplysia kurodai*, and their structures elucidated using inverse-detected NMR methods (Kigoshi et al. 1990). The authors report using a combination of COLOC and HMBC spectra for long-range connectivities utilized in the structure elucidation of these molecules. Unfortunately, no details of the correlations actually observed in the HMBC spectrum were given in their communication. The structure of aplaminone (**50**) is shown.

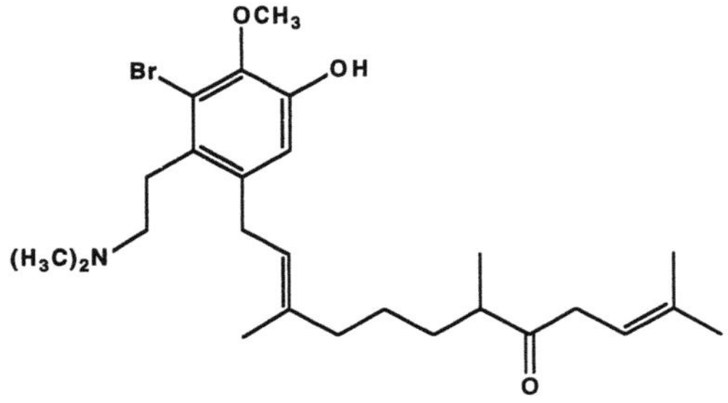

5 0

9.2.10 Guanidine-Derived Alkaloids

A novel polycyclic guanidine alkaloid, ptilomycalin A (**51**), was isolated from a red sponge, *Hemimycale* sp., from the Red Sea (Kashman et al. 1989). The authors report using long-range connectivity information gleaned from both COLOC and HMBC spectra in the elucidation of the structure. Some of the correlations observed in the HMBC spectrum were reported in the paper and are shown on the structure.

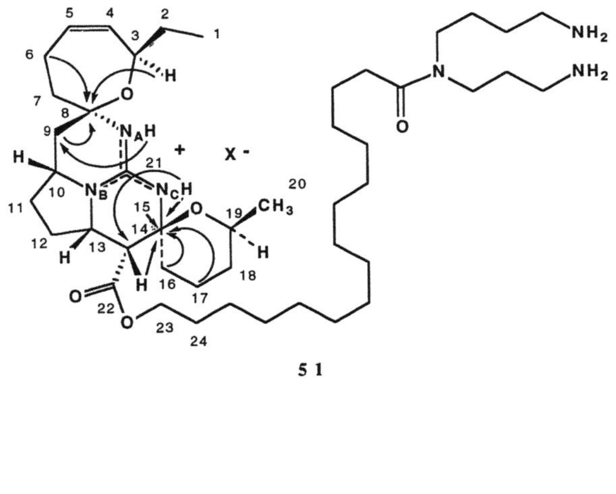

5 1

10 Conclusion

Inverse-detected 2D NMR methods have begun to have a profound impact on the elucidation of alkaloid structures in particular and natural products in general (Martin and Crouch 1991, 1993). The sensitivity of these methods is orders of magnitude higher than their heteronucleus-detected predecessor experiments, allowing structures to be determined on significantly smaller samples. There are an assortment of new techniques, e.g., HMQC-TOCSY and derivative experiments, that provide access to structural information not heretofore available. Some of these experiments have no practical heteronucleus-detected equivalents. Finally, selective one-dimensional and F_1 region-selective analogs of the more familiar two-dimensional NMR experiments are now beginning to appear and can be expected to further expand our experimental horizons.

The complexity of the structural questions that may be probed has also been augmented by inverse-detected experiments. For example, correlations between identical carbon centers in C_2-symmetric molecules may be probed by taking advantage of isotopomeric asymmetry (Kawabata et al. 1992b). In similar fashion, using ^{13}C-coupled HMQC-ROESY data, stereochemical questions may also be probed between identical protons in C_2-symmetric molecules, again using isotopomeric asymmetry (Kawabata et al. 1992a). HMQC-NOESY and -ROESY experiments at natural abundance have rather low sensitivity (Crouch et al. 1990a,b; Kawabata et al. 1992a) relative to even HMBC experiments. It is quite probable that an as-yet-undeveloped selective, one-dimensional analog of either the HMQC-NOESY or -ROESY experiment could be utilized to probe nOe or rOe information for only the carbon of interest. A selective experiment could be performed in a fraction of the time required for a complete 2D experiment, thereby making such experiments practical at natural abundance.

Undoubtedly, even more interesting applications of inverse-detected methods are possible and will be developed as structural problems drive the spectroscopist to find new solutions. Some of the applications that have been presented herein have yet to be applied to an alkaloid. It is, however, only a question of an alkaloid structural problem amenable to solution by a particular technique and an investigator who is aware that the technique exists before corresponding applications in alkaloid chemistry appear. Finally, the recent availability of microinverse and microdual probes (Crouch and Martin 1992a,b) also promises to lead to applications of inverse-detected techniques in the elucidation of the structures of minor alkaloidal constituents that have, until now, been available in quantities too small to allow their structures to be elucidated.

References

Ablordeppy SY, Hufford CD, Bourne RF, Dwuma-Badu D (1990) Proton NMR and carbon-13 NMR assignments of cryptolepine, a 3:4-benzo-d-carboline derivative isolated from *Cryptolepis sanguinolenta*. Planta Med 56:416–417

Alam M, Mroue M (1991) Crooksiine, a bisindole alkaloid from *Haplophyton crooksii*. Phytochemistry 30:1741–1744

Atta-ur-Rahman, Nighat F, Nelofer A, Zaman K, Choudhary MI, DeSilva KTD (1991) Macroxine – a novel oxindole alkaloid from *Alstonia macrophylla*. Tetrahedron 47:3129–3136

Bax A (1983a) Two-dimensional heteronuclear relayed coherence transfer spectroscopy. J Magn Reson 53:149–153

Bax A (1983b) A simple method for the calibration of the decoupler radiofrequency field strength. J Magn Reson 52:76–80

Bax A, Davis DG (1985) MLEV-17 Based two-dimensional homonuclear magnetization transfer spectroscopy. J Magn Reson 65:355–360

Bax A, Drobny G (1985) Optimization of two-dimensional homonuclear relayed coherence transfer NMR spectroscopy. J Magn Reson 61:306–320

Bax A, Marion D (1988) Improved resolution and sensitivity in ^{1}H-detected heteronuclear multiple-bond correlation spectroscopy. J Magn Reson 78:186–191

Bax A, Subramanian S (1986) Sensitivity-enhanced two-dimensional heteronuclear chemical shift correlation NMR spectroscopy. J Magn Reson 67:565–569

Bax A, Summers MF (1986) ^{1}H and ^{13}C Assignments from sensitivity-enhanced detection of heteronuclear multiple bond connectivity by 2D multiple-quantum NMR. J Am Chem Soc 108:2093–2094

Bax A, Freeman R, Kempsell SP (1980) Natural abundance ^{13}C–^{13}C coupling observed via double-quantum coherence. J Am Chem Soc 102:4849–4851

Bax A, Griffey RH, Hawkins BL (1983a) Correlation of proton and nitrogen-15 chemical shifts by multiple-quantum NMR. J Magn Reson 55:301–315

Bax A, Griffey RG, Hawkins BL (1983b) Sensitivity-enhanced correlation of ^{15}N and ^{1}H chemical shifts in natural-abundance samples via multiple-quantum coherence. J Am Chem Soc 105:7188–7190

Bax A, Davis DG, Sarkar SK (1985a) An improved method for two-dimensional heteronuclear relayed coherence transfer NMR spectroscopy. J Magn Reson 63:230–234

Bax A, Ferretti JA, Nashed N, Jerina DM (1985b) Complete ^{1}H and ^{13}C NMR assignment of complex polycyclic aromatic hydrocarbons. J Org Chem 50:3029–3034

Bendall MR, Pegg DT, Doddrell DM (1983) Pulse sequences utilizing the correlated motion of coupled heteronuclei in the transverse plane of the doubly rotating frame. J Magn Reson 52:81–117

Berger S (1988) Selective inverse correlation of ^{13}C and ^{1}H NMR signals, an alternative to 2D NMR. J Magn Reson 81:561–564

Bermel W, Wagner K, Griesinger C (1989) Proton-detected C,H correlation via long-range couplings with soft pulses; determination of coupling constants. J Magn Reson 83:223–232

Bisset NG, Bosly J, Das BC, Spiteller G (1975) Alkaloids from *Strychnos henningsii*. Revised structures for holstiine and rindline, proposed structure for holstiline. Phytochemistry 14:1411–1414

Bodenhausen G, Ruben DJ (1980) Natural abundance nitrogen-15 NMR by enhanced heteronuclear spectroscopy. Chem Phys Lett 69:185–189

Bolton PH (1982) Assignments and structural information via relayed coherence transfer spectroscopy. J Magn Reson 48:336–340

Bolton PH (1985) Heteronuclear relay transfer spectroscopy with proton detection. J Magn Reson 62:143–146

Bolton PH, Bodenhausen G (1982) Relayed coherence transfer spectroscopy of heteronuclear systems: detection of remote nuclei in NMR. Chem Phys Lett 89:139–144

Bosly J (1951) Akaloids of *Strychnos holstii* var. *reticulata*. J Pharm Belg 6:135–149
Braunschweiler L, Ernst RR (1983) Coherence transfer by isostropic mixing: application to proton correlation spectroscopy. J Magn Reson 53:521–528
Brühwiler D, Wagner G (1986) Selective excitation of ^{1}H resonance coupled to ^{13}C. Hetero-COSY and RELAY experiments with ^{1}H detection for a protein. J Magn Reson 69:546–551
Carmely S, Ilan M, Kashman Y (1989) 2-Amino imidazole alkaloids from the marine sponge *Leucetta chagosensis*. Tetrahedron 45:2193–2200
Carmely S, Moore RE, Patterson GML, Corbett TH, Valeriote FA (1990) Tantazoles: unusual cytotoxic alkaloids from the blue-green alga *Scytonema mirabile*. J Am Chem Soc 112:8195–8197
Carroll AR, Scheuer PJ (1990) Kuanoniamines A, B, C, and D: pentacyclic alkaloids from a tunicate and its prosobranch mollusck predator *Chelynotus semperi*. J Org Chem 55:4426–4431
Carroll AR, Cooray NM, Poiner A, Scheuer PJ (1989) A second shermilamine alkaloid from a tunicate *Trididemnum* sp. J Org Chem 54:4231–4232
Cheng J-f, Ohizumi, Wälchli MR, Nakamura H, Hirata Y, Sasaki T, Kobayashi J (1988) Prianosins B, C, and D, novel sulfur-containing alkaloids with potent antineoplastic activity from the Okinawan marine sponge *Prianos melanos*. J Org Chem 53: 4621–4624
Cherif A, Martin GE, Soltero LR, Massiot G (1990) Configuration and total assignment of the ^{1}H and ^{13}C-NMR spectra of the alkaloid holstiine. J Nat Prod 53:793–802
Copp BR, Blunt JW, Munro MHG (1989) A biologically active 1,2,3-trithiane derivative from the New Zealand ascidian *Aplidium* sp. D. Tetrahedron Lett 30:3703–3706
Copp BR, Ireland CM, Barrows LR (1991) Wakayin: a novel cytotoxic pyrroloiminoquinone alkaloid from the ascidian *Clavelina* species. J Org Chem 56:4596–4597
Crouch RC, Martin GE (1991) Selective inverse multiple bond analysis. A simple 1D experiment for the measurement of long-range heteronuclear coupling constants. J Magn Reson 92:189–194
Crouch RC, Martin GE (1992a) Micro inverse-detection: a powerful technique for natural product structure elucidation. J Nat Prod 55:1343–1347
Crouch RC, Martin GE (1992b) Comparative evaluation of conventional 5 mm inverse and micro inverse-detection probes at 500 MHz. Magn Reson Chem 30:S66–S70
Crouch RC, Andrews CW, Martin GE, Luo J-K, Castle RN (1990a) HMQC-NOESY: application to a polynuclear aromatic at natural abundance. Magn Reson Chem 28:774–778 .
Crouch RC, McFadyen RB, Daluge SM, Martin GE (1990b) Disentangling coupling and nOe pathways involving poorly resolved proton signals: HMQC-TOCSY and HMQC-NOESY. Magn Reson Chem 28:792–796
Crouch RC, Shockcor JP, Martin GE (1990c) 1D HMQC-TOCSY: a selective one-dimensional analogue of HMQC-TOCSY. Tetrahedron Lett 31:5273–5276
Crouch RC, Spitzer TD, Martin GE (1992a) Region-selective inverse-detected long-range heteronuclear chemical shift correlation using shaped pulses. Magn Reson Chem 30:595–605
Crouch RC, Spitzer TD, Martin GE (1992b) Strategies for the phase-editing of relayed responses in 2D HMQC-TOCSY spectra. Magn Reson Chem 30:S71–S73
Davis DG (1989) Proton detection of long-range couplings to carbon-13 using semiselective pulses. A high-resolution method for assigning amino acids in peptides. J Magn Reson 83:212–218
Davis DG, Bax A (1985a) Assignment of complex ^{1}H NMR spectra via two-dimensional homonuclear Hartmann-Hahn spectroscopy. J Am Chem Soc 107:2820–2821
Davis DG, Bax A (1985b) Identification of ^{1}H NMR spectra by selective excitation of experimental subspectra. J Am Chem Soc 107:7197–7198
De Bruyn A, Zhang W, Budesinsky M (1989) NMR Study of three heteroyohimbine derivatives from *Rauwolfia serpentina*: stereochemical aspects of the two isomers of reserpiline hydrochloride. Magn Reson Chem 27:935–940

Doddrell DM, Pegg DT, Bendall MR (1982) Distortionless enhancement of NMR signals by polarization transfer. J Magn Reson 48:323–327

Domke T (1991) A new method to distinguish between direct and remote signals in proton-relayed X,H correlations. J Magn Reson 95:174–177

Domke T, Xu P, Freeman R (1991) Geminal-filtered correlation spectroscopy. J Magn Reson 92:218–225

Dwuma-Badu D, Ayim JSK, Fiagbe NIY, Knapp JE, Schiff PL Jr, Slatkin DJ (1978) Constituents of West African medicinal plants XX: quindoline from *Cryptolepis sanguinolenta*. J Pharm Sci 67:433–434

Edwards MW, Daly JW, Myers CW (1988) Akaloids from a panamanian poison frog, *Dendrobates speciosus*: identification of pumiliotoxin-A and allopumiliotoxin class alkaloids, 3,5-disubstituted indolizidines, 5-substituted 8-methylindolizidines, and a 2-methyl-6-nonyl-4-hydroxypiperidine. J Nat Prod 51:1188–1197

Emsley L, Bodenhausen G (1989) Self-refocusing effect of 270° Gaussian pulses. Applications to selective two-dimensional exchange spectroscopy. J Magn Reson 82: 211–221

Fanso-Free SNY, Furst GT, Srinivasan PR, Lichter RL, Nelson RB, Panetta JA, Gribble GW (1979) Organic structure characterization by natural-abundance nitrogen-15 nuclear magnetic resonance spectroscopy. *Rauwolfia* alkaloids and model compounds. J Am Chem Soc 101:1549–1553

Frey MH, Wagner G, Vasak M, Sørensen OW, Neuhaus D, Wörgötter, Kägi JHR, Ernst RR, Wüthrich K (1985) Polypeptide-metal cluster connectivities in metallothionein 2 by novel proton-cadium 113 heteronuclear two-dimensional NMR experiments. J Am Chem Soc 107:6847–6851

Friedrich J, Davies S, Freeman R (1987) Shaped selective pulses for coherence-transfer experiments. J Magn Reson 75:390–395

Fusetani N, Yasumuro K, Matsunaga S, Hirota H (1989) Haliclamines A and B, cytotoxic macrocyclic alkaloids from a sponge of the genus *Haliclona*. Tetrahedron Lett 30: 6891–6894

Garbow JR, Weitekamp DP, Pines A (1982) Bilinear rotation decoupling of homonuclear scalar interactions. Chem Phys Lett 93:504–509

Garnier J, Mahuteau J, Plat M, Merienne C (1989) The complete assignment of the ^{13}C and ^{1}H NMR spectra of jadiffine. Phytochemistry 28:308–309

Geen H, Freeman R (1991) Band-selective radiofrequency pulses. J Magn Reson 93: 93–141

Gramain J-C, Remuson R, Vallee-Goyet D, Guilhem J, Lavaud C (1991) Total synthesis and determination of structure of the pyrrolizidine alkaloid curassanecine. J Nat Prod 54:1062–1067

Gunasekera SP, Gunasekera M, McCarthy P (1991) Discodermide: a new bioactive macrocyclic lactam from the marine sponge *Discodermia dissoluta*. J Org Chem 56:4830–4833

Gunawardana GP, Kohmoto S, Burres NS (1989) New cytotoxic acridine alkaloids from two deep water marine sponges of the family Pachastrellidae. Tetrahedron Lett 30:4359–4362

Hayakawa Y, Kanamaru N, Morisaki N, Furihata K, Seto H (1991a) Lydicamycin, a new antibiotic of a novel skeletal type. J Antibiot 45:288–292

Hayakawa Y, Kanamaru N, Morisaki N, Seto H, Furihata K (1991b) Structure of lydicamycin, a new antibiotic of a novel skeletal type. Tetrahedron Lett 32:213–216

Hugel G, Royer D, Sigaut F, Levy J (1991) Flow thermolysis rearrangements in the indole alkaloid series: 1,2-dehydroaspidospermidine. J Org Chem 56:4631–4636

Isogai A, Fukuchi N, Hayashi M, Kamada H, Harada H, Suzuki A (1990) Mikimopine, an opine in hairy roots of tobacco induced by *Agrobacterium rhizogenes*. Phytochemistry 29:3131–3134

Kashman Y, Hirsh S, McConnell OJ, Ohtani I, Kusumi T, Kakisawa H (1989) Ptilomycalin A: a novel polycyclic guanidine alkaloid of marine origin. J Am Chem Soc 111: 8925–8926

Kato S, Shindo K, Kataoka Y, Yamagishi Y, Mochizuki J (1991) Studies on free radical scavenging substances from microorganisms. II. Neocarazostatins A, B, and C, novel free radical scavengers. J Antibiot 44:903–907

Kawabata J, Fukushi E, Mizutani J (1992a) 2D ^{13}C-coupled HMQC-ROESY: a probe for NOE's between equivalent protons. J Am Chem Soc 114:1115–1117

Kawabata J, Fukushi E, Hara M, Mizutani J (1992b) Detection of connectivity between equivalent carbons in a C_2 molecule using isotopomeric assymmetry: identification of hopeaphenol in *Carex pumila*. Magn Reson Chem 30:6–10

Kay LE, Ikura M, Tschudin R, Bax A (1990) Three-dimensional triple-resonance NMR spectroscopy of isotopically enriched proteins. J Magn Reson 89:496–514

Keniry MA, Poulton GA (1991) Assignment of quaternary carbon resonances in lambertellin by soft heteronuclear multiple bond correlation. Magn Reson Chem 29:46–48

Kessler H, Bernd M, Kogler H, Zarbock J, Sørensen OW, Bodenhausen G, Ernst RR (1983) Peptide conformations. 28. Relayed heteronuclear correlation spectroscopy and conformational analysis of cyclic hexapeptides containing the active sequence of somatostatin. J Am Chem Soc 105:6944–6952

Kessler H, Anders U, Gemmecker G, Steuernagel S (1989a) Improvement of NMR experiments by employing semiselective half-Gaussian-shaped pulses. J Magn Reson 85:1–14

Kessler H, Schmieder P, Kurz M (1989b) Implementation of the DEPT sequence in inverse shift correlation; the DEPT-HMQC. J Magn Reson 85:400–405

Kessler H, Schmieder P, Köck M, Kurz M (1990) Improved resolution in proton-detected heteronuclear long-range correlation. J Magn Reson 88:615–618

Kessler H, Matter H, Gemmecker G, Kling A, Kottenhahn M (1991a) Solution structure of a synthetic N-glycosylated cyclic hexapeptide determined by NMR spectroscopy and MD calculations. J Am Chem Soc 113:7550–7563

Kessler H, Mronga S, Müller G, Moroder L, Huber R (1991b) Conformational analysis of a IgG1 hinge peptide derivative in solution determined by NMR spectroscopy and refined by restrained molecular dynamics simulations. Biopolymers 31:1189–1204

Kessler H, Mronga S, Gemmecker G (1991c) Multi-dimensional NMR experiments using selective pulses. Magn Reson Chem 29:527–557

Kigoshi H, Imamura Y, Yoshikawa K, Yamada K (1990) Three new cytotoxic alkaloids, aplaminone, neoaplaminone and neoaplaminone sulfate from the marine mollusc *Aplysia kurodai*. Tetrahedron Lett 31:4911–4914

Kobayashi J, Cheng J-f, Ishibashi M, Nakamura H, Ohizumi Y, Hirata Y, Sasaki T, Lu H, Clardy J (1987) Prianosin A, a novel antileukemic alkaloid from the Okinawan marine sponge *Prianos melanos*. Tetrahedron Lett 28:4939–4942

Kobayashi J, Cheng J-f, Wälchli MR, Nakamura H, Hirata Y, Sasaki T, Ohizumi Y (1988) Cystodytins A, B, and C, novel tetracyclic aromatic alkaloids with potent antineoplastic activity from the Okinawan tunicate *Cystodytes dellechiajei*. J Org Chem 53:1800–1804

Kobayashi J, Murayama T, Ohizumi Y, Sasaki T, Ohta T, Nozoe S (1989) Theonelladins A–D, novel antineoplastic pyridine alkaloids from the Okinawan marine sponge *Theonella swinhoei*. Tetrahedron Lett 30:4833–4836

Kobayashi J, Cheng J-f, Ohta T, Nozoe S, Ohizumi Y, Sasaki T (1990a) Eudistomidins B, C and D: novel antileukemic alkaloids from the Okinawan marine tunicate *Eudistoma glaucus*. J Org,Chem 55:3666–3670

Kobayashi J, Murayama T, Ishibashi M, Kosuge S, Takamatsu M, Ohizumi Y, Kobayashi H, Ohta T, Nozoe S, Sasaki T (1990b) Hyrtiosins A and B, new indole alkaloids from the Okinawan marine sponge *Hyrtios erecta*. Tetrahedron 46:7699–7702

Kobayashi J, Murayama T, Kosuge S, Kanda F, Ishibashi M, Kobayashi H, Ohizumi Y, Ohta T, Nozoe S, Sasaki T (1990c) Niphatesines A–D, new antineopolastic pyridine alkaloids from the Okinawan marine sponge *Niphates* sp. J Chem Soc Perkin Trans I:3301–3303

Kobayashi J, Cheng J-f, Ishibashi M, Wälchli MR, Yamamura S, Ohizumi Y (1991a) Penaresidin A and B, two novel azetidine alkaloids with potent actomyosin ATPase-activating activity from the Okinawan marine sponge *Penares* sp. J Chem Soc Perkin Trans 1:1135–1137

Kobayashi J, Kanda F, Ishibashi M, Shigemori H (1991b) Manzacidins A–C novel tetrahydropyrimidine alkaloids from the Okinawan marine sponge *Hymeniacidon* sp. J Org Chem 56:4574–4576

Kobayashi J, Tsuda M, Agemi K, Shigemori H, Ishibashi M, Sasaki T, Mikami Y (1991c) Purealidins B and C, new bromotyrosine alkaloids from the Okinawan marine sponge *Psammaplysilla purea*. Tetrahedron 47:6617–6622

Kogler H, Sørensen OW, Bodenhausen G, Ernst RR (1983) Low-pass J Filters. Suppression of neighbor peaks in heteronuclear relayed correlation spectra. J Magn Reson 55:157–163

Lerner L, Bax A (1986) Sensitivity-enhanced two-dimensional heteronuclear relayed coherence transfer NMR spectroscopy. J Magn Reson 69:375–380

Leupin W, Wagner G, Denny WA, Wüthrich K (1987) Assignment of the carbon-13 nuclear magnetic resonance spectrum of a short DNA-duplex with proton-detected two-dimensional heteronuclear correlation spectroscopy. Nucl Acid Res 15:267–275

Levy GC, Lichter RL (1979) Nitrogen-15 nuclear magnetic resonance spectroscopy. John Wiley, New York

Live D, Davis DG, Agosta WC, Cowburn D (1984) Observation of 1000-fold enhancement of ^{15}N NMR via proton-detected multiquantum coherences: studies of large peptides. J Am Chem Soc 106:6104–6105

Live DH, Armitage IM, Dalgarno DC, Cowburn D (1985) Two-dimensional ^{1}H–^{113}Cd chemical-shift correlation maps by ^{1}H-detected multiple-quantum NMR in metal complexes and metalloproteins. J Am Chem Soc 107:1775–1777

Martin GE, Crouch RC (1991) Inverse-detected two-dimensional NMR methods: applications in natural products chemistry. J Nat Prod 54:1–70

Martin GE, Crouch RC (1994) Two-dimensional NMR experiments in natural and unnatural products chemistry. In: Croasmun WR, Carlson RMK (eds) Two-dimensional NMR spectroscopy – applications for chemists and biochemists, chap 11, 2nd edn. VCH, New York, pp 873–914

Martin GE, Spitzer TD, Crouch RC, Luo J-K, Castle RN (1992) Inverted and suppressed direct response HMQC-TOCSY spectra – a convenient method of spectral editing. J Heterocycl Chem 29:577–582

Martin GE, Zektzer AS (1988) Long-range two-dimensionalear heteronuclear chemical shift correlation. Magn Reson Chem 26:631–652

Massiot G, Nuzillard JM, Richard B, Le Men-Olivier L (1990) Alternative partial synthesis of bisindole alkaloids. Tetrahedron Lett 31:2883–2884

Meksuriyen D, Cordell GA (1988) Biosynthesis of staurosporine. 1. ^{1}H and ^{13}C NMR assingments. J Nat Prod 51:884–892

Mukherjee R, da Silva BA, Das BC, Keifer PA, Shoolery JN (1991) Structure and stereochemistry of divaricine, a new bisindole alkaloid from *Strychnos divaricans* Ducke. Heterocycles 32:985–990

Müller L (1979) Sensitivity enhanced detection of weak nuclei using heteronuclear multiple-quantum coherence. J Am Chem Soc 101:4481–4484

Mueller L, Schiksnis RA, Opella SJ (1986) Proton-detected natural-abundance ^{15}N NMR spectroscopy utilizing constant-time multiple-quantum excitation. J Magn Reson 66:379–384

Neuhaus D, Keeler J, Freeman R (1985) Investigation of individual proton spin multiplets by C–H correlation spectroscopy. J Magn Reson 61:553–558

Otvos JD, Engseth HR, Wehrli S (1985) Multiple-quantum ^{113}Cd–^{1}H correlation spectroscopy as a probe of metal coordination environments in metalloproteins. J Magn Reson 61:579–584

Rinehart KL, Holt TG, Fregeau NL, Stroh JG, Keifer PA, Sun F, Li LH, Martin DG (1990) Ecteinascidins 729, 743, 745, 759A, 759B, and 770: potent antitumor agents from the Caribbean tunicate *Ecteinascidin turbinata*. J Org Chem 55:4512–4515

Seebach D, Ko SY, Kessler H, Köck M, Reggelin M, Schmieder P, Walkinshaw MD, Bölsterli JJ, Bevec D (1991) 185. Thiocyclosporins: preparation, solution and crystal structure, and immunosuppressive activity. Helv Chim Acta 74:1953–1990

Sklenar V, Bax A (1987) Two-dimensional heteronuclear chemical-shift correlation in proteins at natural abundance ^{15}N and ^{13}C levels. J Magn Reson 71:379–383

Sohn K, Opella SJ (1989) Determination of ^{1}H homonuclear nOe between amide sites in protins with ^{1}H/^{15}N heteronuclear correlation spectroscopy. J Magn Reson 82:193–197

Spitzer T, Kuehne ME, Bornmann W, Anklin C (1989) Heteronuclear correlation via spin-locked zero- and multiple-quantum coherences. J Magn Reson 84:654–657

Spitzer TD, Crouch RC, Martin GE, Sharaf MHM, Schiff PL Jr, Tackie AN, Boye GL (1991) Total assignment of the proton and carbon NMR specta of the alkaloid quindoline – utilization of HMQC-TOCSY to indirectly establish protonated carbon – protonated carbon connectivities. J Heterocycl Chem 28:2065–2070

Spitzer TD, Crouch RC, Martin GE (1992) Total assignment of the proton and carbon NMR spectra of 5'-NOR-anhydrovinblastine. J Heterocycl Chem 29:265–273

Tackie AN, Sharaf MHM, Schiff PL Jr, Boye GL, Crouch RC, Martin GE (1991) Assignment of the proton and carbon NMR spectra of the indoloquinoline alkaloid cryptolepine. J Heterocycl Chem 28:1429–1435

Tackie AN, Boye GL, Sharaf MHM, Shiff PL Jr, Crouch RC, Spitzer TD, Johnson RL, Dunn J, Minick D, Martin GE (1993) Cryptospirolepine: a unique spiro-nonacyclic alkaloid isolated from *Cryptolepis sanguinolenta*. J Nat Prod 56:653–670

Takayama H, Sun J-J, Aimi N, Sakai S (1989) Lassiocarpine, a novel C_{20}-diterpene alkaloid isolated from *Aconitum kojimae* Ohwi. Tetrahedron Lett 30:3441–3442

Thomas DM, Bendall MR, Pegg DT, Doddrell DM, Field J (1981) Two-dimensional ^{13}C–^{1}H polarization transfer J spectroscopy. J Magn Reson 42:298–306

Ubillas R, Barnes CL, Gracz H, Rottinghaus GE, Tempesta MS (1989) X-ray crystal structure of oxalicine A, a novel alkaloid from *Penicillium oxalicum*. J Chem Soc Chem Commun 21:1618–1619

von Philipsborn W, Müller R (1986) ^{15}N-NMR spectroscopy – new methods and applications. Angew Chem Int Ed Engl 25:383–413

Wilde JA, Boldton PH, Stolowich NJ, Geralt JA (1986) A method for the observation of selected proton NMR resonances of proteins. J Magn Reson 68:168–171

Williamson DS, Smith RA, Nagel DL, Cohen SM (1989) Phase-sensitive heteronuclear multiple-bond correlation in the presence of modest homonuclear coupling. Application to distamycin A. J Magn Reson 82:605–612

Witanowski M, Stefaniak L, Webb GA (1993) Nitrogen NMR spectroscopy. In: Webb GA (ed) Annual reports on NMR spectroscopy. Academic Press, New York

Wright AE, Forleo DA, Gunawardana GP, Gunasekera SP, Koehn FE, McConnell OJ (1990) Antitumor tetrahydroisoquinoline alkaloids from the colonial ascidian *Ecteinascidia turbinata*. J Org Chem 55:4508–4512

Zektzer A, Sims LD, Castle RN, Martin GE (1988) Assignment of the NMR spectra of benzo[b]triphenyleno[1,2-d]thiophene through the application of zero-quantum proton-proton correlation combined with proton-detected long-range multiple quantum heteronuclear chemical shift correlation. Magn Reson Chem 26:287–295

Zuiderweg ERP (1990) A proton-detected heteronuclear chemical-shift correlation experiment with improved resolution and sensitivity. J Magn Reson 86:346–357

Electrochemical Detection of Alkaloids in HPLC

V.-P. RANTA, J.C. CALLAWAY, and T. NAARANLAHTI

1 Introduction

High-performance liquid chromatography (HPLC) is a valuable tool in plant analysis, due to its high power of resolution and compatibility for the determination of nonvolatile and thermally labile compounds. Therefore HPLC has often been applied in the determination of alkaloids (Verpoorte et al. 1984; Popl et al. 1990).

When analyzing complex samples by HPLC, the selection of detection system is important. If chromatographic separation is incomplete or analytic concentration very low, then the more universal and insensitive ultraviolet absorbance (UV) detectors are not satisfactory, and other detection systems must be used. Fluorescense (FL) detection is a sensitive and selective alternative for those compounds that fluoresce. Electrochemical (EC) detection differs from UV and FL in that it is based on a chemical reaction rather than a physical phenomenon, and is the best choice for the many electroactive compounds. In this system analytes are either oxidized or reduced on the electrode surface. More precisely, this technique may be called amperometric or voltammetric detection, though in practice it is commonly referred to as EC.

The breakthrough in liquid chromatography with EC detection (LCEC) soon followed the development of high-efficiency liquid chromatography in the early 1970s. In 1973 Kissinger and co-workers (Kissinger et al. 1973) published their successful account of combining EC detection with HPLC in the determination of catecholamines. This has remained the most popular application of LCEC, though the range of potential applications has expanded greatly during the years. However, the determination of alkaloids by LCEC remains relatively unexplored. The major exception is in the determination of morphine (and codeine to a lesser extent) in body fluids. This application alone covers nearly half of the almost 150 publications concerned with the LCEC detection of alkaloids. Of the remaining half, less than 20 deal with the determination of alkaloids in plants. Perhaps one of the reasons for this neglect is that alkaloids are not often thought of as being electroactive, like phenolic compounds, or that EC is still considered novel and unreliable. This may have been the case over 10 years ago, but today we see that EC is the most sensitive and selective detection system available for many compounds, with typical detection limits in the picogram-nanogram

Modern Methods of Plant Analysis, Volume 15
Alkaloids (ed. by Linskens/Jackson)
© Springer-Verlag Berlin Heidelberg 1994

range, and deserves more attention in the determination of alkaloids in plants. In addition, EC is suitable for many aliphatic alkaloids which cannot be detected by UV or FL without derivatization.

Herein are described the principles, practice, and promise of LCEC in the determination of alkaloids in plant material. First the basic principles of LCEC relative to the determination of alkaloids are explained. Then the most recent methods applicable to LCEC detection of alkaloids in plant material are described, including our own work with *Catharanthus* alkaloids in cell cultures. Finally, we suggest applications for other alkaloids, which are based on published accounts of LCEC detection of alkaloids in body fluids and in other nonplant matrixes.

2 Principles of HPLC with Electrochemical Detection

2.1 Basic Principles of Electrochemical Detection

EC detection is based on an electrochemical surface reaction. When sufficient positive or negative potential is applied to the working electrode of the flow cell, the analyte is either oxidized or reduced on the electrode surface (Fig. 1). The resulting current, most often followed by keeping the potential constant, is directly proportional to the concentration of the analyte passing through the cell. This technique is called amperometry. In general, the lower the oxidation or reduction potential of the analyte, the better the detection limit and selectivity. The potential of the working electrode is controlled with respect to the potential of a reference electrode,

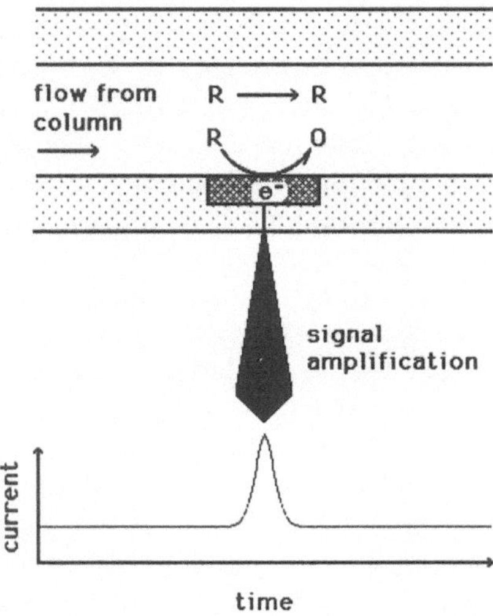

Fig. 1. The electrochemical detector measures the current from the oxidation of the analyte (loss of electrons) which is displayed as a function of time in the chromatogram

which remains at the same potential throughout the measurement. The type of reference electrode used must always be expressed with the working potential. The most common reference electrode is silver/silver chloride (Ag/AgCl). (For additional information, see Shoup 1986; Štulík and Pacáková 1987.)

Since detection is based on a surface reaction, the oxidation or reduction rate of the analyte is dependent on the catalytic properties of the working electrode's surface. Therefore different electrode materials may be selected for the detection of different compounds. On the other hand, the surface reaction may also become a source of problems. The most common is fouling of the working electrode by the analytes or impurities present in the sample or eluent. However, this problem can be managed by either cleaning the electrode or recalibrating the detector.

2.2 Electroactive Functional Groups in Alkaloids

EC is most often used in the analysis of catecholamines and aromatic amines, since these compounds are easily oxidized at low potentials (Table 1). However, most alkaloids also contain oxidizable functional groups, and are well suited for oxidative EC detection. Many contain either a phenol group or an indole nucleus, and even more contain a tertiary aliphatic amine. In addition, many aliphatic alcohols and amines, which are oxidized at high potentials on carbon electrodes, can be detected at much lower potentials with gold or platinum electrodes (Sect. 2.4). Alkaloids, however, do not usually contain easily reducible groups like quinones or aromatic nitro groups.

2.3 Mobile Phase

2.3.1 Conductivity

In EC detection, the mobile phase must conduct the current. Most mobile phases used in reversed-phase and ion-exchange chromatography are capable

Table 1. Oxidation potentials of important functional groups using carbon electrodes

Amines and indoles	Potential (V vs. Ag/AgCl)	Phenols and phenolic ethers	Potential (V vs. Ag/AgCl)
Aromatic amines	+0.6– +1.0	Catechols	+0.4– +0.7
Hydroxyindoles	+0.6– +0.8	Hydroquinones	+0.4– +0.7
Indoles	+0.8– +1.0	Methoxyphenols	+0.7– +0.9
Tertiary aliphatic amines	+0.9– +1.2	Phenols	+0.8– +1.1
Secondary aliphatic amines	>+1.2	Aromatic methoxy groups	>+1.2

of fullfilling this requirement, since they consist of an aqueous buffer and one, or more, polar organic solvents such as methanol or acetonitrile. Such eluents have also been used with unmodified silica columns to a lesser extent. Conductivity is usually adequate when the ionic strength of the buffer is between 0.01–0.1 M.

2.3.2 Background Current and Noise

Background current usually results from the oxidation or reduction of the mobile phase. Since noise increases with background current, EC detection becomes limited at very high potentials. In aqueous solutions, the positive potential limit is essentially restricted by the oxidation of water:

$$2H_2O \rightarrow O_2 + 4H^+ + 4e^-$$

Since this reaction generates hydronium ions, the oxidation potential decreases with increasing pH. The Nernst equation predicts a 59 mV decrease per pH unit, when the reaction generates equal amounts of electrons and hydronium ions. At pH 4.5 the positive potential limit is about $+1.2$ V vs. Ag/AgCl.

One must keep in mind that impurities in the mobile phase contribute to background current. Impurities may also adsorb to the electrode surface. In order to minimize these problems, one should use reagents and solvents of highest purity and keep the buffer concentration low. In addition, the mobile phase additives should be chosen carefully. For example, tertiary aliphatic amines should be avoided as competing bases since they may be oxidized at low potential (Sect. 3.2).

Since the background current is affected by the composition of the mobile phase, isocratic separations are most often employed. However, gradient elution is possible, especially when the applied potential is low ($<+0.8$ V vs. Ag/AgCl) (Khaledi and Dorsey 1985). Thermostating, at least the column, is recommended, since temperature affects both the magnitude of the background current and response of the analyte. In addition, noise can be reduced by using a pulse-free pump and a pulse dampener.

In reductive EC detection, dissolved oxygen must be carefully removed from the mobile phase, as it is reduced to hydrogen peroxide at low negative potential (Kissinger 1989). This makes reductive detection more laborious than oxidative detection. After removal of dissolved oxygen from the mobile phase, the negative potential limit is ultimately restricted by the reduction of hydronium ions to hydrogen.

2.3.3 pH

Care is needed in selecting the mobile phase pH, since it affects separation of the components, EC properties of the analytes, and background current.

When detection is based on the oxidation of a tertiary aliphatic amine, for example, proper pH is essential. Ideally, the nitrogen atom should not be protonated, since the lone pair of electrons is involved in the oxidation process (Schwartz and David 1985). In addition, an unprotonated nitrogen is less reactive towards residual silanol groups on the reversed-phase column, which then improves the peak shape (so-called ion suppression). In order to keep the nitrogen unprotonated, the pH of the mobile phase should be over the pK_a value of the amine. However, the pK_a in eluents containing organic solvents may not always be equivalent to values determined from pure aqueous solutions. Also, the pH must not be higher than necessary, since it does not further the amine's oxidation, but instead promotes the oxidation of water. The choice of pH should always be determined experimentally. One should also consider the stability of the column. Often one must make some compromise between separation and detection.

2.4 Electrode Materials and Flow Cells

The most popular electrode materials are glassy carbon and porous graphite, since they have a large potential range (about $-1.0\,V$ to $+1.2\,V$ vs. Ag/AgCl) with good chemical and mechanical resistance.

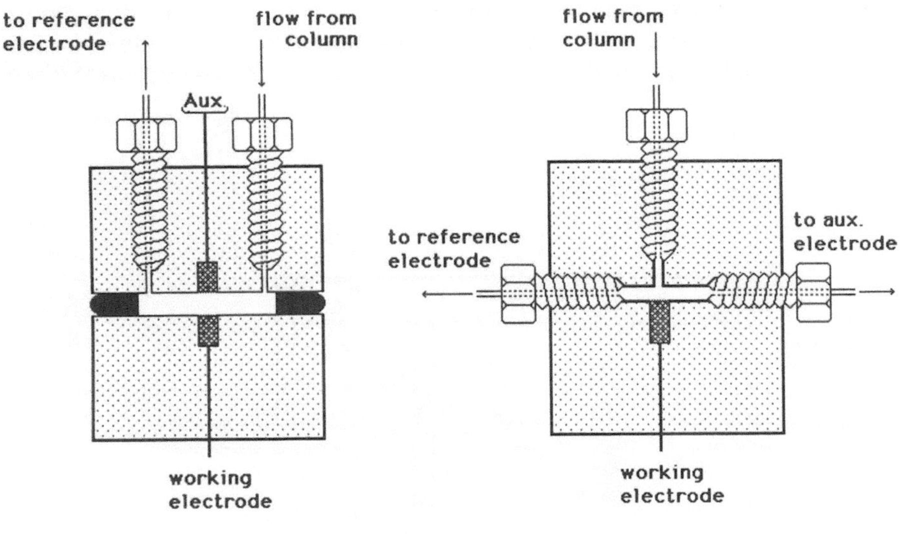

Fig. 2. Amperometric flow cells

Discs of glassy carbon (typically 3 mm in diameter) are most often installed in either thin-layer or wall-jet cells (Fig. 2), with thin-layer being the more popular. A third electrode in the cells, the auxiliary or counter electrode, helps to maintain a well-defined working electrode potential. Thin-layer and wall-jet cells, which are available from several manufacturers, are called amperometric, since they oxidize or reduce a small proportion (typically 1–10%) of the analyte passing through the cell (amperometry is also equated with constant-potential detection). These cells have many practical virtues. The working electrode can be removed for polishing, and replacement if necessary, and dead volumes of the cells can be easily adjusted to less than 1 μl. These cells can also be connected downstream of conventional spectroscopic detectors. In addition, the best thin-layer cells can be equipped with two working electrodes in parallel or series (Sects. 4.1 and 4.2).

Porous graphite is the raw material of flow-through electrodes. Since porous electrodes have a very large surface area, their electrolytic efficiency is much higher than that of disc electrodes. When the electrode oxidizes or reduces nearly all of the analyte, it is referred to as coulometric. The Coulochem Model 5010 cell from Environmental Sciences Associates (ESA) contains two coulometric electrodes made of porous graphite (Fig. 3), the total volume of this cell is about 5 μl. ESA cells are equipped with a solid palladium reference electrode (Pd), with a working potential typically 0.2 V lower than that for Ag/AgCl. The detection limits of coulometric cells are equal to or better than most amperometric cells, and the large porous electrodes have other advantages. The first coulometric electrode in a dual-electrode cell can actually clean the sample more effectively than corre-

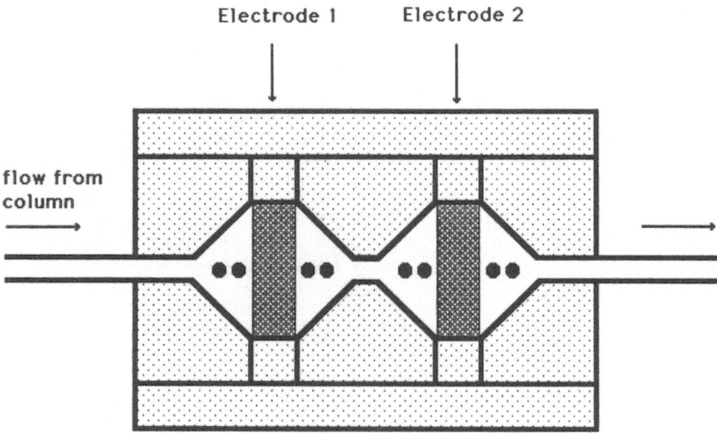

Fig. 3. Dual-electrode coulometric cell; *black dots* represent both reference and auxiliary (counter) electrodes

sponding amperometric electrodes ("screen mode", Sect. 3.2), and the large electrodes are not contaminated as easily as the smaller disc electrodes. On the other hand, porous electrodes cannot be replaced. They are cleaned by flushing the cell with a nitric acid solution and organic solvents. One should also take care in connecting coulometric cells after spectroscopic detectors, as the backpressure resulting from porous electrodes may harm optical flow cells.

Gold and platinum have a high catalytic activity, and therefore many aliphatic alcohols and amines are oxidized on gold or platinum electrodes at much lower potentials than on carbon-based electrodes. On the other hand, gold and platinum electrodes have a tendency to adsorb oxidation products to their surface, which results to a rapid loss of activity. However, Johnson's group found in the early 1980s that the activity could be restored during analysis by potential pulses (Johnson and LaCourse 1990). This method, which is called pulsed amperometric detection (PAD), was soon commercialized. During the last few years, gold electrodes have gained more in popularity than platinum. The flow cell is usually a common thin-layer cell, but more complicated electronics are needed to perform the desired potential waveform (Sect. 3.5). Due to the changes in potential, the detection limits are higher than in constant-potential detection. Today, PAD is very popular in the determination of carbohydrates (including amino sugars) (Johnson and LaCourse 1990) and it has also been applied to the determination of amino acids (Johnson and LaCourse 1990; Martens and Frankenberger 1992). Due to the aliphatic structures of these compounds, they cannot be detected by UV or FL without derivatization. The most popular arrangements for oxidative EC are summarized in Table 2.

2.5 Selection of Applied Potential

The EC properties of an analyte can be studied by measuring its response at different potentials, where the potential is changed incrementally between injections. It is best to start at a high potential and decrease it, since the background current stabilizes more quickly when lowering the potential.

Table 2. Arrangements for oxidative EC detection

Electrode material	Measuring technique	Typical analytes	Detection limits
Glassy carbon	Amperometry	Phenols, aromatic amines, heterocyclic compounds	pg-ng
Porous graphite	"Coulometry"	As above	pg-ng
Gold	Pulsed amperometry	Carbohydrates	ng

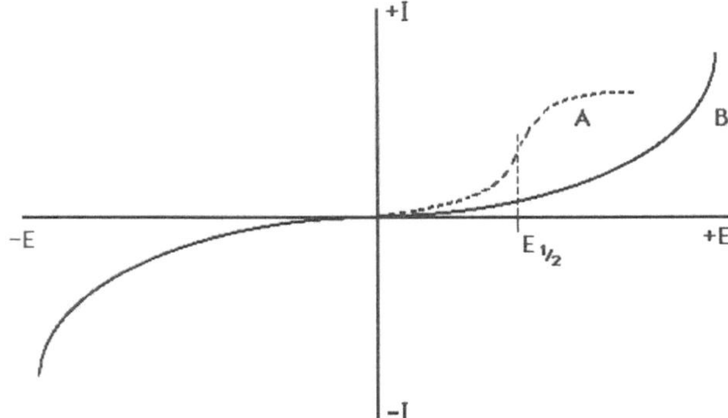

Fig. 4. A hydrodynamic voltammogram of an oxidizable analyte (*A*). The half-wave potential of the analyte ($E_{1/2}$) is the potential where the current is one half of the maximum current. Background current (*B*) results from the oxidation or reduction of the mobile phase. The Y-axis represents current (*I*), usually as micro- or nanoamps, while the X-axis represents potential in volts (+*E* is positive electrode potential)

The resulting curve, where the peak height (or area) is presented as a function of potential, is called a hydrodynamic voltammogram (Fig. 4). This can be considered as an EC spectrum of the analyte. The half-wave potential is characteristic of the analyte under these conditions, and is independent of concentration. The EC properties can also be studied by scanning the potential (voltammetry). However, in the final optimization a hydrodynamic voltammogram created by detection at a constant potential is usually required.

In selecting the applied potential, one should consider the signal-to-noise (S/N) ratio, reliability, and selectivity. One should also measure the magnitude of the background current and noise when creating the voltammogram, then the S/N ratio can be expressed as a function of potential. When the analyte is oxidized at low potential, the S/N ratio is usually highest at the plateau of the voltammogram. Choosing an applied potential from the plateau often ensures good reliability, since small changes in the potential do not affect the response of the analyte. However, there are situations when one should consider choosing a potential from the rising portion of the voltammogram. When the analyte is oxidized at a higher potential, then the S/N ratio may be highest near the half-wave potential, since at higher potentials the background current and noise increase faster than the response of the analyte. In addition, when an interfering compound is oxidized at a slightly higher potential than the analyte, the selectivity of the measurement may be improved by choosing a potential nearer the half-wave potential of the analyte.

3 Methods of Detection in Plants

The order of the methods described in this section is from easily oxidizable alkaloids to those which are more difficult to oxidize. Specified conditions of these methods are presented in Table 3.

3.1 Psilocin and Psilocybin in Fungi

Determination of the indole alkaloids psilocin and psilocybin in fungi has been the most popular application of EC in the analysis of alkaloids in a plant matrix. This research has primarily been done by two groups: Christiansen and co-workers in Norway, and Kysilka and Wurst in the Czech Republic. Especially psilocin is well suited for EC, due to its hydroxy substituent on the indole ring, essentially a phenol, whereas in psilocybin this group is phosphorylated. Since the pH of the mobile phase is not critical to the oxidation of either the phenolic group or indole nucleus, it has been possible to use different separation techniques.

Christiansen et al. developed an HPLC method of screening Norwegian mushrooms for the presence of psilocin and psilocybin. The alkaloids were extracted from powdered dried mushrooms with methanol which contained 10% 1 N ammonium nitrate (Christiansen et al. 1981). The recovery of psilocybin was at least 98%, whereas the recovery of psilocin was not reported. The analytes were separated on an unmodified silica column using an alkaline eluent (Christiansen and Rasmussen 1983). After separation, the effluent was split to UV and FL detectors, and an EC detector was connected downstream of the UV detector. Psilocybin was identified and determined by both UV and FL detection. EC was optimized for psilocin due to its low concentration in the mushroom. The detection limit of psilocin by EC was 100 and 250 times lower than by UV and FL, respectively. Psilocybin gave only a small response by EC, since it is optimally oxidized at a higher potential. By this method, it was found that *Conocybe cyanopus* and *Pluteus salicinus* also contained psilocin and psilocybin (Christiansen et al. 1984).

Kysilka and Wurst have studied the biosynthesis of psilocin and psilocybin in mushrooms for many years. Recently, they thoroughly studied the extraction of psilocin and psilocybin and found that different extraction systems must be used for each alkaloid (Kysilka and Wurst 1990). Psilocin was extracted with best recovery by ethanol:water (3:1) and psilocybin by methanol:water (3:1) saturated with potassium nitrate. The recoveries of psilocin and psilocybin were about 10 and 1.3 times higher, respectively, than with just methanol extraction for *Psilocybe bohemica*. These researchers also used a separation technique different from that of Christiansen's group. Psilocin and psilocybin were eluted on a reversed-

Table 3. Recent LCEC methods for alkaloids in plant matrixes

Analyte	Matrix	Column	Mobile phase	Cell[a]	Potential[b]		Detection limit[c] (ng)	Reference
					E_{pre}	E_{det}		
Psilocin Psilocybin	Fungi	Partisil 5 (5 µm) 250 × 4.6 mm	73:23:3 MeOH:H_2O: 1 M NH_4NO_3 (pH 9.6); 1 mM Na_2EDTA	TL		+0.65	0.11 7.5	Christiansen and Rasmussen (1983)
Psilocin Psilocybin	Fungi	Silasorb SPH C18 (7.5 µm) 250 × 4 mm	95:5:1 H_2O:MeOH: Acetic acid	WJ		+1.2	0.7 1.0	Kysilka (1990); Wurst et al. (1992)
Hordenine	Germinating barley	Nucleosil SA (5 µm) 200 × 4.6 mm	80:20 NaH_2PO_4 (ionic strength 0.1, pH 3.0): MeOH	CO-screen	+0.50	+0.75	1.1	Johansson and Schubert (1990)
Ajmalicine Catharanthine	Cell culture	Supelcosil LC-18-DB (5 µm) 150 × 4.6 mm	49:36:15 0.1 M NaAc (pH 6.5):ACN:MeOH	CO-screen	+0.20	+0.80	0.22 0.25	Naaranlahti et al. (1989)
Atropine	Belladonna preparations	LiChrosorb Diol (7 µm) 250 × 4 mm	80:20 0.0125 M sodium phosphate (pH 7.2):ACN	TL		+1.2[d]	2.9[e]	Leroy and Nicolas (1987)
Castanospermine	Leaves	Dionex CS3	0.01 M HCl; post-column addition of 0.3 M NaOH	PAD		+0.05	0.75	Donaldson et al. (1990)

[a] TL, thin-layer cell equipped with a glassy carbon electrode; WJ, wall-jet cell equipped with a glassy carbon electrode; CO-screen coulometric cell with dual-electrodes of porous graphite used in screen mode; PAD, pulsed amperometric detector equipped with a gold electrode.
[b] E_{pre}, potential of a preanalytical electrode in volts; E_{det}, potential of a detecting electrode in volts. Potentials of coulometric cells versus Pd, potentials of other cells versus Ag/AgCl.
[c] Amount of the analyte injected with a S/N ratio of 3. [d] Potential versus SCE. [e] The lower limit of linear dynamic range.

phase column using an acidic eluent (Kysilka 1990; Wurst et al. 1992). The detection system consisted of a UV photodiode array detector and an EC detector connected in series. Since quite a high potential was applied, a high response was obtained for both psilocin and psilocybin. EC was found to be 30 and 20 times more sensitive than UV for psilocin and psilocybin, respectively. However, only psilocin was routinely quantified by EC. By this method, levels of psilocin and psilocybin were determined from the fruiting bodies of *Psilocybe* and *Inocybe*, and from mycelial extracts of *Psilocybe bohemica*.

3.2 Hordenine in Germinating Barley

Hordenine [4-(N,N-dimethylaminoethyl)phenol] is structurally related to tyramine, as well as several other compounds, such as catecholamines. It was earlier used in both human and veterinary medicine for the treatment of diarrhea and other intestinal disturbances.

Johansson and Schubert (1990) analyzed hordenine from germinating barley. Samples were first incubated in 1 M hydrochloric acid at 95 °C for 30 min, then homogenized and centrifuged. The supernatant was diluted with the mobile phase before injection. The researchers first tried to separate hordenine from its precursors, tyramine and N-methyltyramine, on a reversed-phase column using an acidic eluent. However, there was significant tailing of the peaks. Apparently the protonated amines reacted with residual silanol groups of the stationary phase. The peak shape could be improved by adding a competing base (N,N-dimethyl-N-octylamine) to the eluent. Unfortunately, the retention of the analytes decreased and separation was inadequate. Addition of an ion-pairing reagent (octylsulfate) increased the retention, but separation was still not satisfactory. In addition, the competing base started to oxidize at a low potential, leading to a high background current. When the reversed-phase column was replaced by a strong cation-exchange column, good separation and peak symmetry resulted without additives.

Hordenine was detected with a dual-electrode coulometric cell. The potential of the first electrode was set to +0.50 V [vs. solid palladium reference electrode (Pd)], which was at the base of hordenine's hydro-dynamic voltammogram. This electrode cleaned the sample by removing easily oxidizable impurities. Hordenine and its precursors were subsequently detected by a second electrode at a potential of +0.75 V. This method of detection is called the "screen mode". In addition, the mobile phase was purified by connecting a coulometric cell between the pump and injector as a guard cell. This additional cell operated at +0.80 V. The detection limit of hordenine was at 1.1 ng, which was 25 times better than detection by UV at 275 nm.

3.3 *Catharanthus* Alkaloids in Cell Culture

The Madagascan periwinkle, *Catharanthus roseus* (L.) G. Don, produces over 70 different indole alkaloids, many of which have pharmacological activity. The most important alkaloids are vinblastine and vincristine, which are used as antineoplastic agents. Due to the high demand for these alkaloids, we have studied the possibility of producing them in cell cultures. However, such cultures do not represent the complete biosynthetic potential of the intact plant, therefore only intermediates were obtained and in yields much lower than from plants. Biosynthetically these intermediates are formed through the combination of tryptophan with different configurations of the monoterpenoid skeletons (Fig. 5). Vinblastine and vincristine are formed in plants by joining vindoline with catharanthine. For selecting high-producing plants for cell culture, we used HPLC with UV detection (Naaranlahti et al. 1987). For the actual cell culture studies we needed a more sensitive method and therefore utilized EC with a dual-electrode coulometric detector (Naaranlahti et al. 1989).

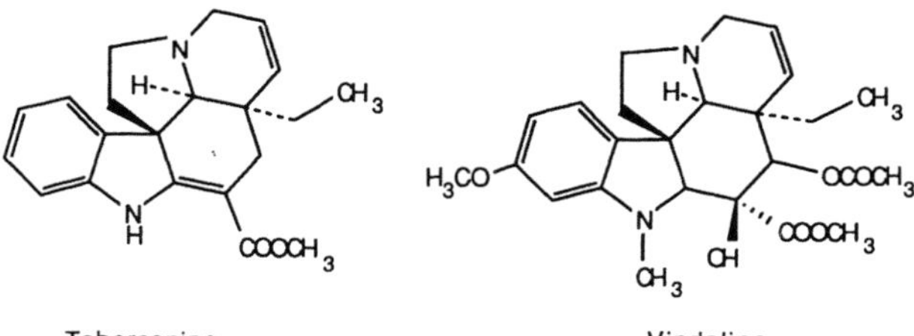

Fig. 5. Structures of the *Catharanthus* alkaloids

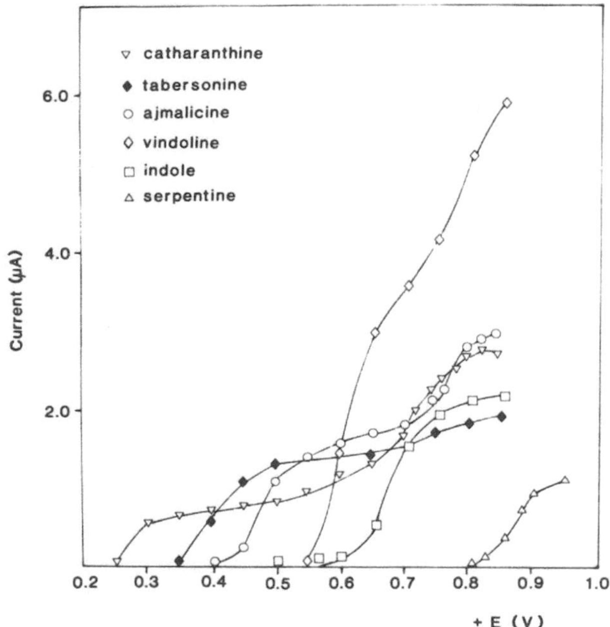

Fig. 6. Hydrodynamic voltammograms of the *Catharanthus* alkaloids. The graphs were constructed by injecting 75 pmol of each compound without column using the chromatographic eluent. Electrode potential versus Pd. (Naaranlahti et al. 1989)

The alkaloids of interest contained both an indole nucleus and a tertiary alicyclic amine. The buffer pH was adjusted to 6.5, since this is very near the pK_a values of the amines; 5.5 for vindoline and 6.8 for catharanthine. This pH also offered good separation between ajmalicine and catharanthine, the major alkaloids in our cell cultures, and ensured good column stability.

The electrochemical properties of the alkaloids were studied at pH 6.5 by creating hydrodynamic voltammograms (Fig. 6). The voltammograms of ajmalicine and catharanthine, which contain an intact indole nucleus and a tertiary aliphatic amine, showed a second clear inflection point at about +0.7 V (vs. Pd), which was near the half-wave potential of indole. These results suggested that the tertiary amine might be oxidized first at the lower potential, followed by oxidation of the indole nucleus at a higher potential, though the potential of the first oxidation of catharanthine was surprisingly low for a tertiary amine. Serpentine, which differs from ajmalicine by having a positively charged quaternary amine and a larger aromatic region, was oxidized at a much higher potential than ajmalicine. Tabersonine and vindoline differ from other *Catharanthus* alkaloids by having an indoline (dihydroindole) nucleus. The indoline nitrogen is secondary in tabersonine and tertiary in vindoline, and they differ substantially in their electro-chemical properties.

Based on these electrochemical studies we developed a method for the quantitation of ajmalicine and catharanthine in cell cultures. These alkaloids were extracted from freeze-dried cells and purified by the solid-phase procedure described by Morris et al. (1985), except that ethanol was used as the extracting solvent instead of methanol. A dual-electrode coulometric cell was used in the screen mode. The potential of the first electode was set at +0.2 V (vs. Pd), which was at the base of catharanthine's voltammogram. The alkaloids were detected by the second electrode at +0.8 V, as this offered the best S/N ratio. Higher potentials led to lower S/N ratio, since the background current and noise started to increase exponentially above +0.85 V, due to the oxidation of water. The mobile phase was purified by a guard cell between the pump and injector. The guard cell operated at +0.8 V.

The detection limits of ajmalicine and catharanthine were 0.22 ng and 0.25 ng, respectively. This method was also selective, since usually no interfering peaks were observed in the chromatograms (Fig. 7). To insure reliability, a two-point calibration was performed daily, since the response varied, especially when starting after a weekend. Therefore it is recom-

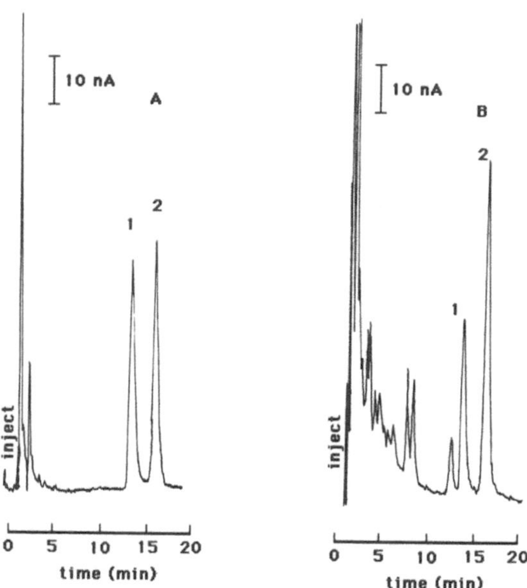

Fig. 7. A Chromatogram of a standard mixture containing 4.2 ng of catharanthine (*1*) and 3.5 ng of ajmalicine (*2*). **B** Chromatogram of a *Catharanthus roseus* cell suspension extract. For details see text and Table 3. (Naaranlahti et al. 1989)

mended to keep the system going at a low flow rate. In addition to calibration, the porous electrodes were cleaned approximately once a month by flushing with a nitric acid solution. Since we used a guard cell, we recycled the eluent for 2 or 3 days, and let the new eluent stabilize overnight. The guard cell had to be cleaned after several months of use, as contamination eventually resulted in increased backpressure. However, these inconveniences were compensated by the low detection limits.

3.4 Atropine in *Belladonna* Preparations

Leroy and Nicolas (1987) analyzed atropine in a powder and tincture of *Belladonna*, and in multicomponent tablets containing *Belladonna*. The sample was first dissolved in dilute sulfuric acid and washed with diethyl ether, the aqueous phase was made alkaline (pH 9–11) and the alkaloids were extracted with dichlormethane. Finally, the organic phase was evaporated and the residue was redissolved in acetonitrile.

Because the oxidizable group in atropine is a tertiary nitrogen atom (pK_a 9.8), the response increased as the pH of the mobile phase was increased. A buffer of pH 7.2 was chosen, as it offered a high response and ensured column stability. Atropine and scopolamine were separated on a hydrophilic-diol column. Reversed-phase columns were also tested, but gave long retention times. However, scopolamine was not completely resolved from the early eluting interferences on the diol column and only atropine was quantitated.

The operating potential was chosen near the upper plateau of the voltammogram, since this offered the best compromise between a high response and a low background current (+1.2 V vs. SCE, saturated calomel electrode, similar to +1.2 V vs. Ag/AgCl). The structurally related alkaloids scopolamine and apoatropine were also oxidized at this potential. The lower limit of the linear dynamic range was 2.9 ng for atropine. EC offered at least a tenfold lower detection limit and wider linear response than UV at 220 nm.

3.5 Castanospermine in Leaves

Castanospermine is an indolizidine alkaloid obtained from the leaves of *Castanospermum australe*. It is known to be a potent inhibitor of glucosidase enzymes and possesses antitumor activity. Since it is a structural analog of a simple sugar molecule (Fig. 8), it is difficult to detect. Low volatility hampers separation by gas chromatography and lack of a chromophore makes sensitive HPLC analysis with photic detection impossible.

However Donaldson et al. (1990) detected castanospermine with pulsed amperometric detection, which is a popular method for detecting carbohy-

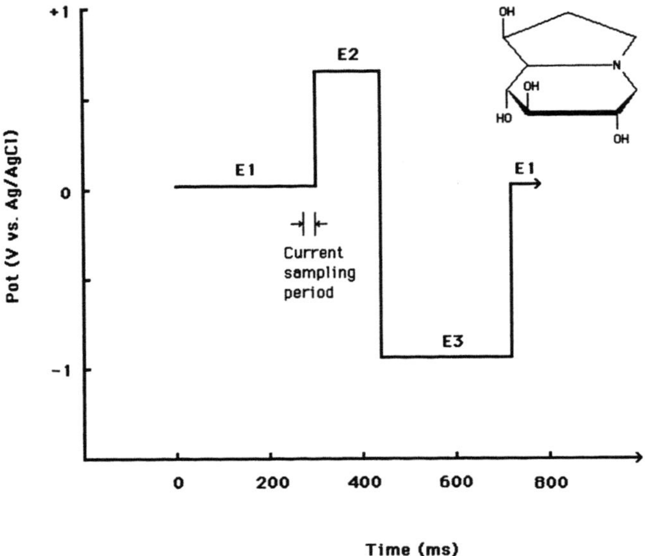

Fig. 8. Pulsed amperometric detection of castanospermine. *E1* (+0.05 V, 300 ms), *E2* (+0.65 V, 120 ms), *E3* (−0.95 V, 300 ms) refer to detection, cleaning and reactivation potentials, respectively

drates. Castanospermine was extracted by simply grinding the leaf material in water with a pestle and mortar, and eluted on a cation exchange column at low pH. Since polyhydroxylated aliphatic compounds are oxidized on a gold electrode only under strongly alkaline conditions, a sodium hydroxide solution was added post-column.

Castanospermine was detected with a triple-pulsed technique, which kept the electrode activity constant (Fig. 8). Castanospermine was first oxidized at a low positive electrode potential (E1), and the current from the oxidation was measured at the end of the pulse, after the disturbing charging current had decayed. Then the electrode surface was cleaned by a high positive potential (E2), which oxidized the electrode surface, leading to desorption of absorbed oxidation products. The surface oxides were subsequently dissolved by a large negative potential (E3) that restored the pure metal surface and thus the activity of the electrode surface. One cycle took less than 1 s.

This method was quite sensitive and selective. The detection limit in aqueous solution was 0.75 ng, and from plant material 2.5 ng of castanospermine could be detected. This method was also found to be suitable for the structurally related alkaloids 1-deoxynojirimycin, 1-deoxymannojirimycin, and swainsonine.

4 Possible Methods for Detecting Other Alkaloids

In this section we discuss the possibilities of using EC for other alkaloids, based on published applications in nonplant matrixes. By careful modification, these methods could be applied to analysis in plants. Representative methods are compiled in Table 4.

4.1 Opium Alkaloids

EC is currently the most popular method for the determination of morphine in body fluids, since morphine contains a phenol group which is easily oxidized (Tagliaro et al. 1989). The detection limit of morphine typically varies from a few tens of picograms to several hundred. Other opiates contain a tertiary alicyclic amine, and can best be detected using a neutral or slightly alkaline eluent and about a 0.3–0.4 V higher potential than for morphine (Schwartz and David 1985). For the nonphenolic alkaloids, codeine has been detected by EC in body fluids at the nanogram level. However, EC has seldom been used for the determination of opium alkaloids in plant material.

Lurie and McGuinness (1987) demonstrated the selectivity of EC for phenolic opiates, when they determined 6-monoacetylmorphine in adulterated heroin samples. The samples were complex and contained several other alkaloids (acetylcodeine, noscapine, and papaverine) with different adulterants such as sugars, procaine, caffeine, and quinine. A thin-layer cell with two working electrodes in parallel was connected after a UV photodiode array detector. In this configuration, each electrode measured the sample independently. The potentials were set to +1.0 V and +1.1 V [vs. silver/silver chloride reference electrode (Ag/AgCl)]. Due to the acidic eluent (pH 2.2), only 6-monoacetylmorphine and papaverine gave a significant response for EC. When heroin was highly adulterated with quinine, 6-monoacetylmorphine was not completely resolved from quinine and could not be quantified with the photodiode array detector. However, 6-monoacetylmorphine could easily be determined by EC, since quinine gave no response. In addition, the parallel electrode system helped to identify 6-monoacetylmorphine and evaluate peak purity, since the response ratio of a sample peak could be compared with a standard.

Recently, it has been demonstrated that improved selectivity for morphine in plasma can be achieved by using a strongly alkaline eluent (pH 9.5–11) (Tagliaro et al. 1989; Jordan and Hart 1991). This is because ionization of the phenol group facilitates oxidation, and hence a very low oxidation potential (+0.35–+0.45 V vs. Ag/AgCl) can be applied; but despite a low potential, the detection limit does not increase. However, a

Table 4. Recent LCEC methods for alkaloids in nonplant matrixes

Analyte	Matrix	Column	Mobile phase	Cell[a]	Potential[b] E_{pre}	E_{det}	Detection limit[c] (ng)	Reference
Cabergoline (ergot derivative)	Plasma Urine	Nucleosil C18 (5 µm) 250 × 4 mm	80:20 0.075 M KH_2PO_4 (pH 3): ACN	CO-screen	+0.35	+0.65	0.03	Pianezzola et al. (1992)
Capsaicin Dihydrocapsaicin	Plasma	Cosmosil 5Ph (5 µm) 150 × 4.6 mm	55:45 0.1 M KH_2PO_4:ACN; pH 5	TL		+0.75	0.012 0.012	Kawada et al. (1985)
Cocaine The methyl ester of ecgonine	Plasma	Hamilton PRP-1 (10 µm) 150 × 4.6 mm	1–8 min 92:8 0.015 M $(NH_4)_3PO_4$: ACN; pH 8.8. 8–20 min 50:50 same components and pH	CO-screen	+0.30	+1.0	10 10	Miller and DeVane (1991)
Homoharringtonine	Plasma	Lichrosorb CN (5 µm) 125 × 4 mm	85:15 0.015 M Na_2HPO_4 (pH 6.8): ACN; 40°C	TL		+1.0	1.2	Chan et al. (1989)
Morphine	Serum	Phase separations S5W-C18	80:20 0.05 M K_2HPO_4 (pH 11):20	TL		+0.45	0.035	Jordan and Hart (1991)
Nicotine	Plasma	µBondapak C18 (10 µm) 300 × 3.9 mm	95:4:1 0.002 M NaH_2PO_4 + 0.25 mM sodium octyl sulfate:ACN: MeOH; pH 3	CO-screen	+0.45	+0.75	0.2	Chien et al. (1988)
Physostigmine	Plasma Urine	Spherisorb S3W (3 µm) 150 × 4.6 mm	45:45:10 ACN: MeOH:0.1 M NH_4NO_3 (pH 8.8)	CO-redox	+0.8 (1) −0.2 (2)	+0.10	0.01	Hurst and Whelpton (1989)

Physostigmine	Plasma	Biophase C8 (5 μm) 250 × 4.6 mm	60:40 0.1 M NaH$_2$PO$_4$ (pH 3.0): ACN; 34 mM sodium dodecylsulphate	TL-redox	+1.0	+0.10	0.07	Isaksson and Kissinger (1987)
Reserpine Rescinnamine	Urine	Biophase ODS (5 μm) 250 × 4.6 mm	65:35 MeOH: 0.05 M KH$_2$PO$_4$ (pH 4.5); 5 mM sodium heptane sulfonic acid	TL		+0.90 +1.0	0.9 0.8	Wang and Bonakdar (1986)
Sparteine	Urine	Spherisorb S5 CN (5 μm) 250 × 4.6 mm	43:32:25 0.01 M NaH$_2$PO$_4$ (pH 2.5): MeOH:ACN	CO-screen	+0.30	+0.90	0.05	Moncrieff (1990)
Vinblastine Vincristine	Plasma Urine	Hypersil ODS (5 μm) 150 × 3.9 mm	65:35 MeOH: 0.01 M Phosphate buffer (pH 7)	WJ		+0.85	0.1 0.1	Vendrig et al. (1988)
Vincamine	Plasma	LiChrosorb RP-8 250 × 4.6 mm	50:50 0.1 M LiClO$_4$: ACN	WJ		+1.2	1.0	Smyth (1986)
Yohimbine	Plasma	Supelcosil C8-DB (5 μm) 250 × 4.6 mm	65:35 H$_2$O:ACN; 0.008 M (C$_2$H$_5$)$_4$NClO$_4$, pH 6.0	CO-screen	+0.25	+0.80	0.1	Hariharan et al. (1991)

[a] CO-screen, coulometric cell with dual-electrodes of porous graphite used in screen mode; CO-redox, coulometric system with three electrodes of porous graphite used in redox mode; TL, thin-layer cell equipped with a glassy carbon electrode; TL-redox, thin-layer cell with dual-electrodes of glassy carbon used in redox mode; WJ, wall-jet cell equipped with a glassy carbon electrode.
[b] E_{pre}, potential of a preanalytical electrode in volts; E_{det}, potential of a detecting electrode in volts. Potentials of coulometic cells versus Pd, potentials of other cells versus Ag/AgCl.
[c] Detection limits may not be exactly comparable since they are not expressed in uniform manners in the original articles. When possible, the detection limit is defined as the amount of the analyte injected with a S/N ratio of 3.

pH-resistant column is required. This principle could propably be used in plant analysis for alkaloids that contain an aromatic hydroxyl group.

4.2 Physostigmine

Hurst and Whelpton (1989) detected physostigmine in plasma and urine using a three-electrode redox system which consisted of a guard cell with one coulometric electrode and an analytical cell with two coulometric electrodes connected in series (Fig. 9). Physostigmine was initially oxidized by the first electrode, then the oxidation product was reduced by the second electrode, and finally the reduction product from the second electrode and any catechol derived from the second chemical reaction were oxidized by the third electrode. This process improved the detection limit, since the background current and noise of the third electrode were lower than that of the first two electrodes, due to a very low positive potential. Physostigmine was eluted on an unmodified silica column using a slightly alkaline eluent. The pH of the eluent was not excessively high, since physostigmine is hydrolyzed to the corresponding phenol, eseroline, under strongly alkaline conditions (pH > 9.5). This method was very sensitive with a detection limit of about 10 pg.

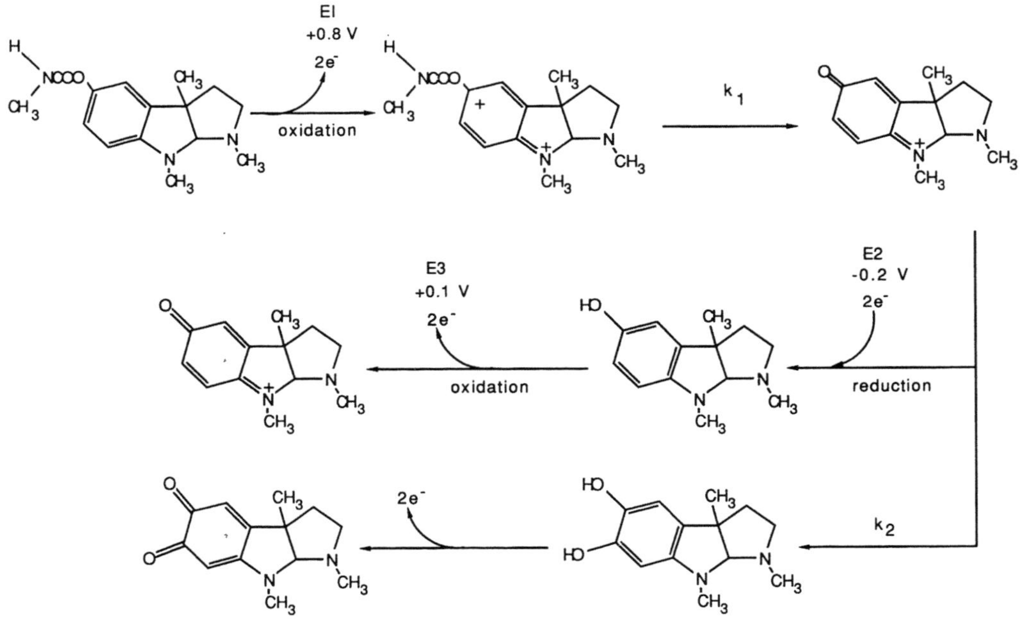

Fig. 9. Redox mode detection mechanism for physostigmine. Electrode potentials versus Pd; k_1 and k_2 represent chemical reaction rates. Detection system from Hurst and Whelpton (1989), structures from Isaksson and Kissinger (1987)

Isaksson and Kissinger (1987) detected physostigmine in plasma with a different system. Physostigmine was eluted on a C-8 column using an acidic eluent with an ion-pairing reagent. This detection system was based on the redox mode, where a dual-electrode thin-layer cell was used in a series configuration, i.e., the reduction current from the second electrode was measured.

4.3 Coca Alkaloids

Schwartz and David (1985) found that the tertiary amine in cocaine and its precursors, ecgonine and benzoylecgonine, could be oxidized by a glassy carbon electrode using a slightly alkaline eluent (pH of buffer was 8) with a high percentage of acetonitrile (50%) as the organic modifier. However, responses of ecgonine and benzoylecgonine were low. Since cocaine was oxidized more easily than its precursors, it could be optimally detected with buffer of pH 7. This small decrease in pH resulted in a lower background current and noise, and better stability of the C-18 column without decreasing the response. Under these conditions, the detection limit of cocaine was 3 ng.

Recently Miller and DeVane (1991) modified the method of Schwartz and David for the determination of cocaine, benzoylecgonine, and the methyl ester of ecgonine in sheep plasma, under basic conditions using a pH-resistant polymer column and a dual-electrode coulometric detector. In addition, separation was improved by using a solvent switching system. Cocaine and the methyl ester of ecgonine could be determined with a detection limit of about 10 ng, whereas benzoylecgonine gave no response. Benzoylecgonine was determined by using an UV detector at 235 nm, connected in series with the coulometric detector. It is worth noting that the methyl ester of ecgonine gave no UV response due to its aliphatic structure.

It seems that ecgonine and benzoylecgonine are more difficult to oxidize than the corresponding methyl esters. It would be interesting to see if better detectability could be obtained by PAD using a gold electrode, since this technique has been applied succesfully in the determination of aliphatic amino acids which are impossible to detect using carbon electrodes (Johnson and LaCourse 1990; Martens and Frankenberger 1992).

4.4 Sparteine

Sparteine is an alkaloid obtained from lupin beans with antiarrhythmic and oxytocic properties. It has a tetracyclic aliphatic structure with two tertiary alicyclic nitrogens. Since sparteine does not contain a chromophore, it cannot be detected with UV or FL. Therefore, Moncrieff (1990) developed a method of detecting sparteine, and its dehydro metabolites, in urine with

EC. She found that these compounds could be oxidized even in an acidic eluent. Satisfactory elution was achieved on a cyano column at pH 2.5. The eluent had to be quite acidic since extensive peak tailing occurred above pH 4.

5 Conclusions and Future Directions

Oxidative EC detection offers advantages over UV and FL for many alkaloids, especially for those that contain a phenol group, indole nucleus or a tertiary aliphatic amine group. Aliphatic alkaloids resembling sugars can be detected with pulsed amperometric detection, and this technique may be applicable to some other aliphatic alkaloids which are difficult or impossible to detect by conventional EC.

For the future, one may expect the development of new electrode materials that will expand the range of applications and make detection more sensitive, selective, and reliable. Also, instead of using new materials, one can improve the catalytic activity or reactivity of existing materials by chemical modification, e.g., adding electron-transfer catalysts or enzymes to their surfaces. The second major trend has been to obtain more information on the sample composition, e.g., connecting multiple electrodes in series or parallel, or installing different types of electrodes in single flow cells. The third trend is universal to analytical chemistry, i.e., to analyze smaller samples with smaller apparatus. The smallest commercial cells are already compatible with microbore columns, and home-made carbon fiber electrodes have been used with capillary columns and in capillary electrophoresis, permitting the analysis of single nerve cells.

Acknowledgment. We would like to thank Seppo Turunen for his help in information retrieval.

References

Chan YP, Lee FW, Siu TS (1989) Quantitation of homoharringtonine in plasma by high-performance liquid chromatography with amperometric detection. J Chromatogr 496:155–166

Chien CY, Diana JN, Crooks PA (1988) High-performance liquid chromatography with electrochemical detection for the determination of nicotine in plasma. J Pharm Sci 77:277–279

Christiansen AL, Rasmussen KE (1983) Screening of hallucinogenic mushrooms with high performance liquid chromatography and multiple detection. J Chromatogr 270:293–299

Christiansen AL, Rasmussen KE, Tønnesen F (1981) Determination of psilocybin in *Psilocybe semilanceata* using high-performance liquid chromatography on a silica column. J Chromatogr 210:163–167

Christiansen AL, Rasmussen KE, Høiland K (1984) Detection of psilocybin and psilocin in Norwegian species of *Pluteus* and *Conocybe*. Planta Med 50:341–343

Donaldson MJ, Broby H, Adlard MW, Bucke C (1990) High pressure liquid chromatography and pulsed amperometric detection of castanospermine and related alkaloids. Phytochem Anal 1:18–21

Hariharan M, Guthrie S, Kindt EK, van Noord T, Grunhaus LJ (1991) Liquid chromatographic coulometric assay and preliminary pharmacokinetics of yohimbine in man. J Liq Chromatogr 14:351–364

Hurst PR, Whelpton R (1989) Solid phase extraction for an improved assay of physostigmine in biological fluids. Biomed Chromatogr 3:226–232

Isaksson K, Kissinger PT (1987) Determination of physostigmine in plasma by liquid chromatography with dual-electrode amperometric detection. J Liq Chromatogr 10:2213–2229

Johansson IM, Schubert B (1990) Separation of hordenine and N-methyl derivatives from germinating barley by liquid chromatography with dual-electrode coulometric detection. J Chromatogr 498:241–247

Johnson DC, LaCourse WR (1990) Liquid chromatography with pulsed electrochemical detection at gold and platinum electrodes. Anal Chem 62:589A–597A

Jordan PH, Hart JP (1991) Voltammetric behaviour of morphine at a glassy carbon electrode and its determination in human serum by liquid chromatography with electrochemical detection under basic conditions. Analyst 116:991–996

Kawada T, Watanabe T, Katsura K, Takami H, Iwai K (1985) Formation and metabolism of pungent principle of *Capsicum* fruits. XV. Microdetermination of capsaicin by high-performance liquid chromatography with electrochemical detection. J Chromatogr 329:99–105

Khaledi MG, Dorsey JG (1985) Hydro-organic and micellar gradient elution liquid chromatography with electrochemical detection. Anal Chem 57:2190–2196

Kissinger PT (1989) Biomedical applications of liquid chromatography-electrochemistry. J Chromatogr 488:31–52

Kissinger PT, Refshauge C, Dreiling R, Adams RN (1973) An electrochemical detector for liquid chromatography with picogram sensitivity. Anal Lett 6:465–477

Kysilka R (1990) Chromatographic determination of psilocybin and psilocin in fruit bodies and mycelia of hallucinogenic mushrooms. Chem Listy 84:988–992 (in Czech)

Kysilka R, Wurst M (1990) A novel extraction procedure for psilocybin and psilocin determination in mushroom samples. Planta Med 56:327–328

Leroy P, Nicolas A (1987) Determination of atropine in pharmaceutical dosage forms containing vegetal preparations, by high-performance liquid chromatography with UV and electrochemical detection. J Pharm Biomed Anal 5:477–484

Lurie IS, McGuinness K (1987) The quantitation of heroin and selected basic impurities via reversed phase HPLC. II. The analysis of adulterated samples. J Liq Chromatogr 10:2189–2204

Martens DA, Frankenberger WT Jr (1992) Pulsed amperometric detection of amino acids separated by anion exchange chromatography. J Liq Chromatogr 15:423–439

Miller RL, DeVane CL (1991) Determination of cocaine, benzoylecgonine and ecgonine methyl ester in plasma by reversed-phase high-performance liquid chromatography. J Chromatogr 570:412–418

Moncrieff J (1990) Simultaneous determination of sparteine and its 2-dehydro and 5-dehydro metabolites in urine by high-performance liquid chromatography with electrochemical detection. J Chromatogr 529:194–200

Morris P, Scragg ÅH, Smart NJ, Stafford A (1985) In: Dixon RA (ed) Plant cell culture; a practical approach. IRL, Oxford, pp 127–167

Naaranlahti T, Nordström M, Huhtikangas A, Lounasmaa M (1987) Determination of *Catharanthus* alkaloids by reversed-phase high-performance liquid chromatography. J Chromatogr 410:488–493

Naaranlahti T, Ranta V-P, Jarho P, Nordström M, Lapinjoki SP (1989) Electrochemical detection of indole alkaloids of *Catharanthus roseus* in high-performance liquid chromatography. Analyst 114:1229–1231

Pianezzola E, Bellotti V, La Croix R, Strolin Benedetti M (1992) Determination of cabergoline in plasma and urine by high-performance liquid chromatography with electrochemical detection. J Chromatogr 574:170–174

Popl M, Fähnrich J, Tatar V (1990) Chromatographic analysis of alkaloids. Marcel Dekker, New York

Schwartz RS, David KO (1985) Liquid chromatography of opium alkaloids, heroin, cocaine and related compounds using electrochemical detection. Anal Chem 57: 1362–1366

Shoup RE (1986) Liquid chromatography/electrochemistry. In: High-performance liquid chromatography, vol 4. Academic Press, New York, pp 91–194

Smyth MR (1986) Determination of vincamine in plasma by high-performance liquid chromatography with voltammetric detection. Analyst 111:851–852

Štulík K, Pacáková V (1987) Electroanalytical measurements in flowing liquids. Ellis Horwood, Chichester

Tagliaro F, Franchi D, Dorizzi R, Marigo M (1989) High-performance liquid chromatographic determination of morphine in biological samples: an overview of separation methods and detection techniques. J Chromatogr 488:215–228

Vendrig DEMM, Teeuwsen J, Holthuis JJM (1988) Analysis of *Vinca* alkaloids in plasma and urine using high-performance liquid chromatography with electrochemical detection. J Chromatogr 424:83–94

Verpoorte R, Baerheim Svendsen A (1984) Chromatography of alkaloids, Part B: gas-liquid chromatography and high-performance liquid chromatography. Elsevier, Amsterdam

Wang J, Bonakdar M (1986) Liquid chromatography of reserpine and rescinnamine using electrochemical detection. J Chromatogr 382:343–348

Wurst M, Kysilka R, Koza T (1992) Analysis and isolation of indole alkaloids of fungi by high-performance liquid chromatography. J Chromatogr 593:201–208

Gas Chromatography in the Analysis of Alkaloids

D. Dagnino and R. Verpoorte

1 Introduction

Since its introduction at the end of the 1950s, gas chromatography (GC) has developed into a versatile tool in the analysis of natural products. Clear advantages of GC are the high resolution and high sensitivity of the most common detection method, the flame ionization detector (FID), and the fact that the detector response of similar compounds will be about the same, i.e., peak areas may be directly compared for quantification. This is in contrast to high performance liquid chromatography (HPLC), in which the detector response in the most common detection mode, ultra violet (UV)-absorption, may vary widely for different compounds since the molar extinction coefficients can be very different.

By using a nitrogen-specific detector (PND) in GC analysis, sensitivity for alkaloids can be even further improved while at the same time selectivity is introduced. The ultimate selectivity is reached by coupling GC with a mass spectrometer. As the gas flow through capillary columns is low, the GC effluent may be introduced directly into the ionization source of a mass spectrometer. This means that both electron impact (EI) and chemical ionization (CI) are possible as ionization mode for the analytes of the GC separation. One may thus obtain fragmentation spectra, both in the positive and negative ion mode. Generally GC-MS (mass spectrometry) will thus provide much more information than LC (liquid chromatography)-MS, where in most types of interfaces only protonated molecules [M+1] (positive ion mode) or deprotonated molecules [M−1] (negative ion mode) are obtained without fragmentation. Also a comparison to MS-MS, where only a separation on mass occurs and consequently no separation of stereo-isomers is possible, is in favor of GC-MS.

Already in 1960 Lloyd et al. showed that alkaloids could be analyzed by means of GC, but the number of applications of GC developed for the analysis of alkaloids remained limited when compared with HPLC (Verpoorte and Baerheim Svendsen 1984; Popl et al. 1990). This is due to the fact that alkaloids are only slightly volatile and the high temperatures necessary to volatilize alkaloids readily cause decomposition. Decomposition has frequently been reported on packed (metal) columns and in injection ports, but various methods have been developed to solve this problem (Verpoorte and Baerheim Svendsen 1984; Popl et al. 1990). With the fused silica capillary columns developed in recent years, the risk of decomposition has further decreased, thus opening new possibilities for the analysis of alkaloids.

Given the apparent advantages of GC, the GC of alkaloids will be reviewed to investigate to what extent this technique is also useful for the analysis of this group of

Modern Methods of Plant Analysis, Volume 15
Alkaloids (ed. by Linskens/Jackson)

compounds. The review will give special emphasis on the coupling with MS. The fragmentation of the alkaloids will not be discussed (for this see References or literature dedicated to the mass spectrometry of alkaloids). Hesse (1974) and Hesse and Bernhard (1975) reviewed this area extensively. For older literature on GC, dealing mainly with packed columns, see the review by Verpoorte and Baerheim Svendsen (1984). A valuable comparison of several chromatographic methods for the analysis of alkaloids of forensic interest was made by Hashimoto et al. (1988).

2 Pyrrolizidine Alkaloids

Pyrrolizidine alkaloids are found in several plant species distributed around the world. They occur mainly in the Boraginaceae, Asteraceae, and Fabaceae. They are thought to have ecological significance as protective chemicals for the plants. Some insects are known to store pyrrolizidine alkaloids for their own chemical defence or to use them as pheromone precursors (Edgar and Culvenor 1974). They are also known to contaminate animal and human food sources and to be present in plants used in phytotherapy (Roeder and Neuberger 1988). Since some of these compounds show carcinogenic and hepatotoxic activities, their analysis has been studied extensively. The pyrrolizidine alkaloids lack a strong chromophore and are thus less suited for HPLC-UV analysis, so that GC seems to be an attractive method for the analysis of these compounds.

GC of pyrrolizidine alkaloids was extensively investigated by Chalmers et al. (1965) using packed columns. They described the analysis of 58 underivatized monocarboxylic acid esters and macrocyclic diester alkaloids and derivatives.

Analysis of *Senecio* pyrrolizidine alkaloids by GC has been reported frequently (Borstel et al. 1989; Pieters et al. 1989; Dimenna et al. 1980; Bicchi et al. 1985, 1989a,b; Hartmann and Toppel 1987; Stelljes et al. 1991; Witte et al. 1992). In *S. vulgaris*, around 90% of the total pyrrolizidine alkaloids were shown to be present as their N-oxides (Hartmann and Toppel 1987). Since the N-oxides are not analyzable by GC usually a reduction of the extract is carried out with Zn dust under acidic conditions. An estimation of the amount of compound present as the N-oxide and as tertiary alkaloid is made by analyzing the alkaloid extract (without reduction) and the extract obtained with the reduction step. Figure 1 illustrates a GC separation of a pyrrolizidine alkaloid extract from *S. inaequidens*. Further examples on the ability of GC to separate complex mixtures of underivatized pyrrolizidine alkaloids are given by Luethy et al. (1983) and Roeder and Neuberger (1988). Metabolites of pyrrolizidine alkaloids have also been analyzed by GC (Ames and Powis 1978; Winter et al. 1988; Mattocks and Jukes 1990).

The efficiency of the quantification of pyrrolizidine alkaloids by GC-FID and GC-PND was compared with quantification made through ^{1}H-NMR

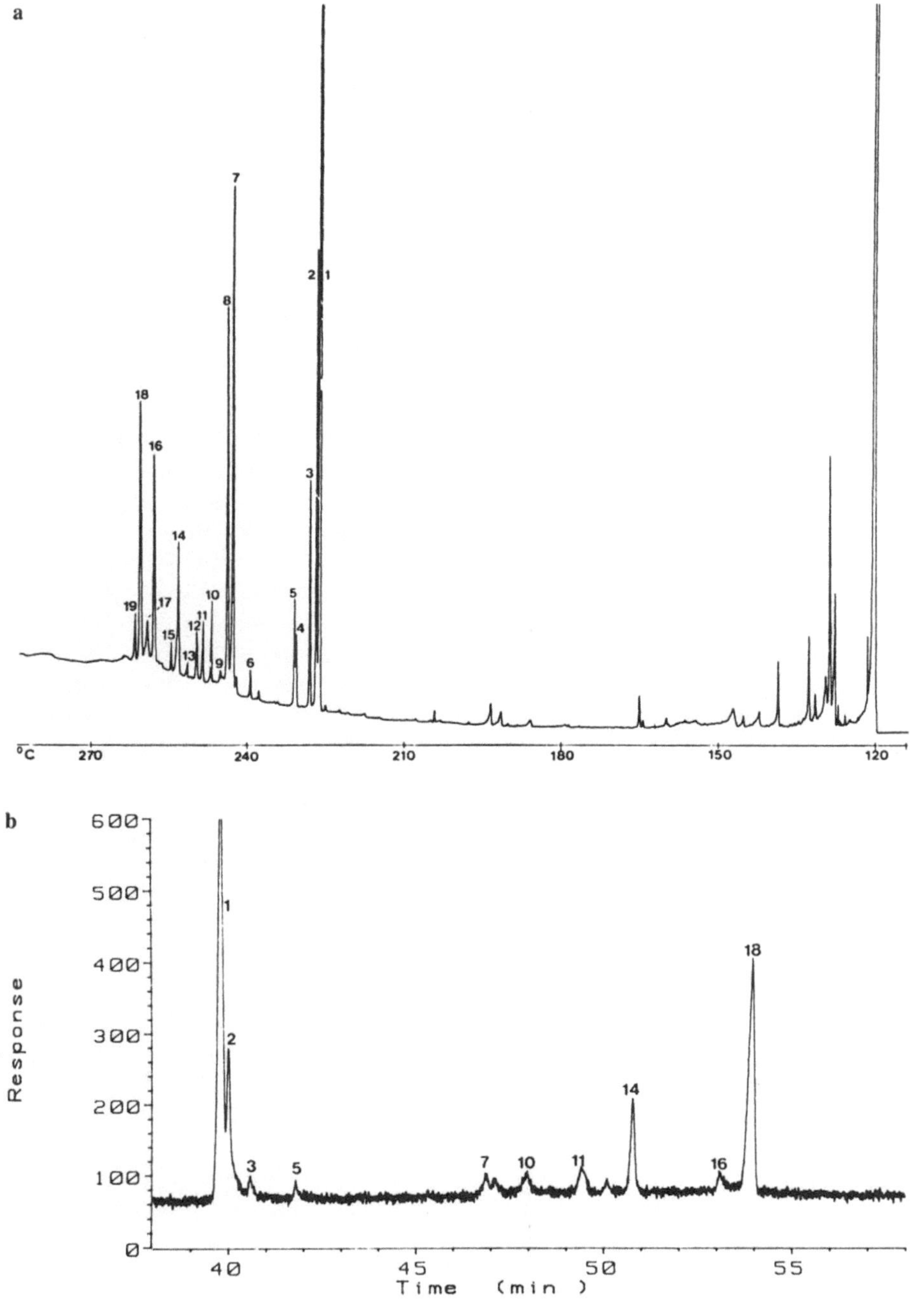

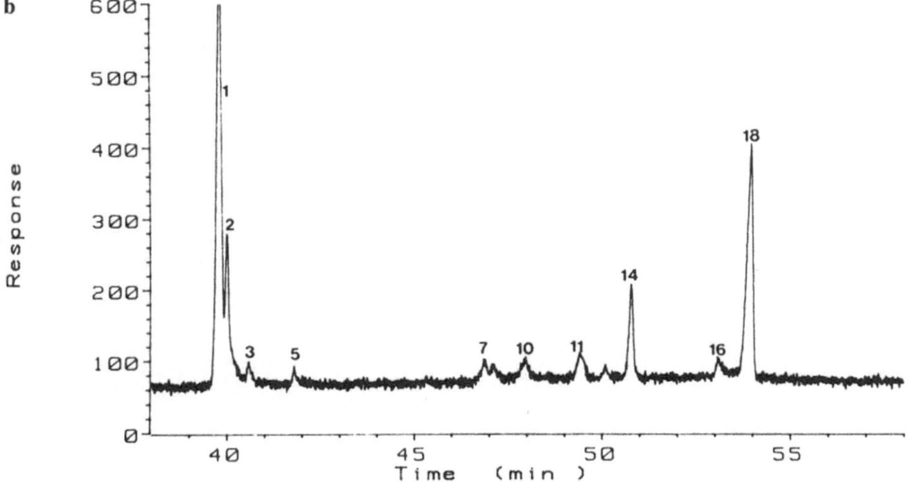

Fig. 1. a Capillary GC-FID pattern of a pyrrolizidine alkaloid extract from *Senecio inaequidens*. **b** Section of GC-FT-IR pattern in which the pyrrolizidine alkaloids eluted. *1* Senecivernine; *2* senecionine; *3* seneciphylline; *4* spartioidine; *5* integerrimine; *6* senkirkine; *7* retrorsine; *9* neo-senkirkine; *10* usaramine; *11* otosenine; *13* O-acetylsenkirkine; *14* desacetyldoronine; *16* florosenine; *17* floridanine; *18* doronine. The GC conditions are described in Table 1. (Bicchi et al. 1989b)

(nuclear magnetic ressonance) and ^{13}C-NMR and the advantages of each of the methods were discussed (Pieters et al. 1989). The investigation showed that GC and ^{1}H-NMR gave similar results for the quantification of total pyrrolizidine alkaloids. For the quantification of the individual alkaloids the main advantage of GC over ^{13}C-NMR is its much higher sensitivity.

Not all pyrrolizidine alkaloids can be analyzed without previous derivatisation. Capillary columns have reduced the necessity of derivatization (Roeder and Neuberger 1988), but this step is still frequently carried out when analyzing monoesters or diesters, e.g., of *Eupatorium* (Hendriks et al. 1987, 1988), *Symphytum* (Huizing et al. 1981; Stengl et al. 1982) and *Amsinckia* species (Dimenna et al. 1980). Most of the GC analyses of pyrrolizidine alkaloids are carried out, either after or without derivatization, on apolar columns. Column length varied considerably, ranging from 15 to 50 m. Temperature programs typically start at 120 °C and go up to 290 or 320 °C when samples have been derivatized. Table 1 lists some examples of GC systems in which pyrrolizidine alkaloids have been separated.

GC has been coupled to a mass spectrometer to provide structural information. EI is mostly used, PICI (positive ion chemical ionization) and NICI (negative ion chemical ionization) were shown to provide additional information for structure elucidation. CI-MS with NH_3 as reactant gas (PICI-MS) resulted in an intense protonated molecule at m/z $[M+1]^+$ (Hendriks et al. 1987, 1988) together with fragment ions that can characterize the necine. OH^- as NICI reagent ion produced either the pseudomolecular ion $[M-H]^-$ or the molecular ion $[M]^-$ and also gives a clue as to the type of necine. NICI-MS with NH_3 as the reactant gas is less informative and presents mostly only a high intensity peak corresponding to either the deprotonated molecule $[M-H]^-$ for the retronecine derivatives or the molecular ion $[M]^-$ for the ontonecine derivatives (Bicchi et al. 1989a).

Table 1. Some GC systems used for the analysis of pyrrolizidine alkaloids

Column	GC conditions				Reference
Type, length (m) × i.d. (mm)	Injector	Temperature program (°C)	Detector	Carrier gas	
WCOT DB-1[a], 25 × 0.25	250° s	120–290° 8°/min	FID PND	He, 0.7 bar	Hartmann and Toppel (1987)
PS 264 (0.3 um), 30 × 0.32	300° s	120° 1 min –280° 3°/min	FID 300 °C	H_2, 3 ml/min	Bicchi et al. (1989b)
CP SIL 5cb[a], 25 × 0.32[b]	300° s	150–300° 6°/min	MS		Hendriks et al. (1987)

s = Split injection.
l = Splitless injection.
[a] Columns with stationary phases of similar polarity. Each can be substituted by any of the others. WCOT, Wall-coated open tubular.
[b] Sample derivatized.

Another method which has been applied to obtain further structural information is GC coupled to a Fourier transform infrared (FT-IR) spectrophotometer. This technique allows the distinction of structural isomers which are otherwise difficult to distinguish by MS (Bicchi et al. 1989b). Since the diastereoisomers can be separated by GC, identification with this method can be carried out on a routine basis once the retention times for the pure stereoisomers have been established (Hendriks et al. 1987).

3 Quinolizidine Alkaloids

Quinolizidine alkaloids are found in several tribes of the Fabaceae. These compounds have been implicated in acute intoxication and death of grazing stock. Since sweet lupin varieties have gained considerable importance as a stock food in some countries, monitoring of quinolizidine alkaloid levels is necessary since high levels may lead to fodder rejection or intoxications.

GC analysis of these compounds was already reported by Lloyd et al. in 1960. Ten compounds were analyzed on a SE-30 packed column. Since then other investigations have been carried out using packed columns analyzing the quinolizidine alkaloid content of *Sophora* (Hatfield et al. 1977; Kinghorn et al. 1982) and *Lupinus* (Keller and Zelenski 1978; Kinghorn et al. 1980; Balandrin and Kinghorn 1981; Keller et al. 1983).

Wink et al. (1980) analyzed by capillary GC the quinolizidine alkaloids from plants and cell suspension cultures of *Lupinus polyphyllus*. Nearly all the alkaloids reported to occur in this species could be detected using this method. Tables with the retention indices of a large number of quinolizidine alkaloids and their characteristic ion fragments (EI-MS) have been given by Wink et al. (1981a, 1983) and Muehlbauer et al. (1988). Using the same method, the analysis of quinolizidine alkaloids from several other genera have been reported by the same group (Wink et al. 1981b, 1982; Wink and Hartmann 1980; Hatzold et al. 1983; Strack et al. 1991).

PND was shown to be 10 to 50 times more sensitive than FID (Wink et al. 1982). The other detection method commonly used to obtain further information on the chemical structures was EI-MS. However, even GC-MS data are not always sufficient for the identification of quinolizidine alkaloids, since structural isomers such as tetrahydrorhombifoline and N-methylangustifoline were found to have the same retention time and very similar mass spectra but could be distinguished by TLC (thin-layer chromatography) (Balandrin and Kinghorn 1981). From some esters of hydroxylupanine it was not possible to detect the molecular ion (Wink et al. 1982). FD-MS (field desorption) was performed directly on the alkaloid extract to detect the molecular ions of the various bases in the mixture. The identity of the compounds was further confirmed by comparison of their retention indices with the ones of authentic samples. More recently, also CI-MS (isobutane or

NH$_3$ as reagent gas) was used to establish the molecular weight (Muehlbauer et al. 1988).

All the capillary GC analyses were carried out without a previous derivatization step on relatively short (15–25 m) apolar columns. The compounds analyzed had a molecular weight ranging from 169 (lupinine) to 394 (13-cinnamoyloxylupanine). Table 2 lists some GC systems used for the analysis of quinolizidine alkaloids. Figure 2 illustrates a GC separation of an alkaloid mixture of *L. polyphyllus*. Their stability at high temperatures and the lack of a strong chromophore makes GC a very attractive method to be applied on a routine basis.

4 Tropane Alkaloids

Tropane alkaloids include some medicinally important secondary plant metabolites as atropine (racemic mixture of l- and d-hyoscyamine) and scopolamine. This group also includes cocaine, which, since it is a major drug of abuse, is frequently analyzed in forensic science. GC analysis of this class of compounds was first shown to be possible by Lloyd et al. in 1960.

Tropane alkaloids from the Solanaceae have been frequently investigated by GC. The BSA (N,O-Bis(trimethylsilyl)acetamide) derivatives of scopolamine and l-hyoscyamine have been quantified in extracts of suspension cultures and redifferentiated roots of *Hyoscyamus niger* (Yamada and Hashimoto 1982; Hashimoto and Yamada 1983) using a packed apolar column. More recently, the tropane alkaloid content of some species of this genus have been analyzed without derivatization with capillary apolar columns (Parr et al. 1990), using a method developed by Witte et al. (1987).

Hartmann et al. (1986) reinvestigated the alkaloid composition of *Atropa belladonna* plants, root cultures and cell suspension cultures. Twenty

Table 2. Some GC systems used for the analysis of quinolizidine alkaloids (for explanations to Footnotes, see Table 1)

Column	GC conditions				Reference
Type, length (m) × i.d. (mm)	Injector	Temperature program (°C)	Detector	Carrier gas	
DB-1[a], 25 × 0.3	250° s	150–320° 10°/min	FID 320°	He, 0.9 bar	Muehlbauer et al. (1988)
WCOT CP SIL 5cb[a], 25 × 0.25	n.m. s	190–270° 6°/min	FID PND	He, 1.2 bar	Hatzold et al. (1983)
OV-101[a], 12 × 0.2	250° s	60° 1 min –120° 30°/min, 0.5 min 120° –260° 10°/min	FID 300°	He, 1.2 bar	Priddis (1983)

n.m. = not mentioned

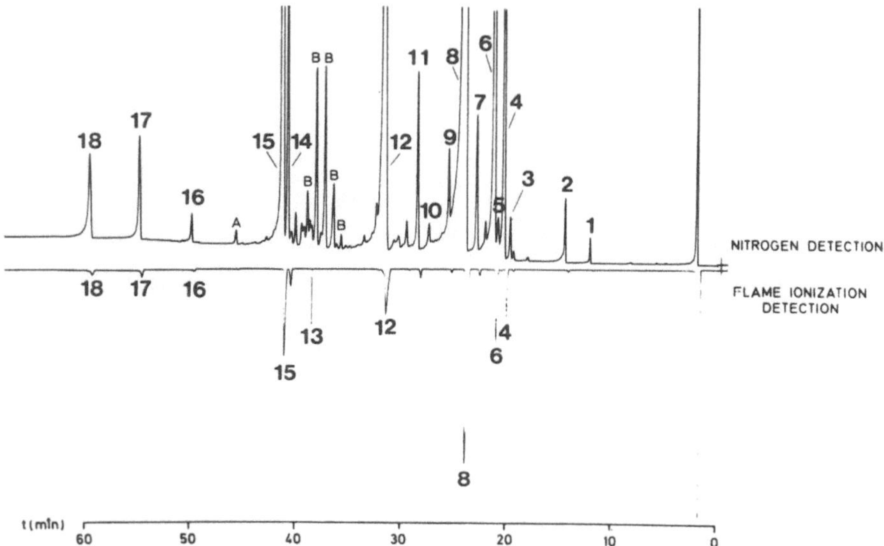

Fig. 2. Separation of an alkaloid mixture from *Lupinus polyphyllus* leaflets obtained by capillary GC. *1* Sparteine; *2* ammodendrine; *3* isoangustifoline; *4* N-methylangustifoline; *5* 17-oxosparteine; *6* angustifoline; *7* dehydrolupanine; *8* lupanine; *9* methylalbine; *10* 4-hydroxylupanine; *11* 17-oxolupamine; *12* 13-hydroxylupanine; *13* 13-(2-methylbutyryl)-oxylupanine; *14* 13-angeloyloxylupanine; *15* 13-tigloyloxylupanine; *16* 13-benzoyloxylupanine; *17* 13-*cis*-cinnamoyl-oxylupanine; *18* 13-*trans*-cinnamoyl-oxylupanine. *A* Unknown ester; *B* containing kresyl phosphates. The separation was carried out in a fused silica capillary column (SE 30, 15 m × 0.23 mm i.d.), temperature program 150–300 °C 6 °/min, at 300 °C isothermal (Wink et al. 1982)

nitrogen-containing compounds were detected with the PND detector and the identity of most of them could be determined by GC-MS. A list of the retention indices of the compounds was presented. The compounds identified include tropane alkaloids and several of their biosynthetic precursors. These precursors were also investigated by Simola et al. (1989). Ylinen et al. (1986) described a method for quantification of atropine and scopolamine in *A. belladonna*.

Several species of *Datura* were analyzed by GC. Witte et al. (1987) investigated the alkaloid pattern of *D. innoxia* plants. With the method used, they were able to detect nearly all alkaloids reported to be present in this species. Littorine was not detected, but that might have been due to the fact that it coelutes with hyoscyamine in the system used. A list of the retention times and the characteristic ions (EI-MS) of 30 tropane alkaloids was included. Apoatropine and aposcopolamine are thought to be artifacts formed during the extraction procedure or in the injector of the GC. Figure 3 illustrates a GC separation of *D. innoxia* root extract.

All the capillary GC analyses of extracts from plant material mentioned above were carried out without a derivatisation step. However, Deutsch et

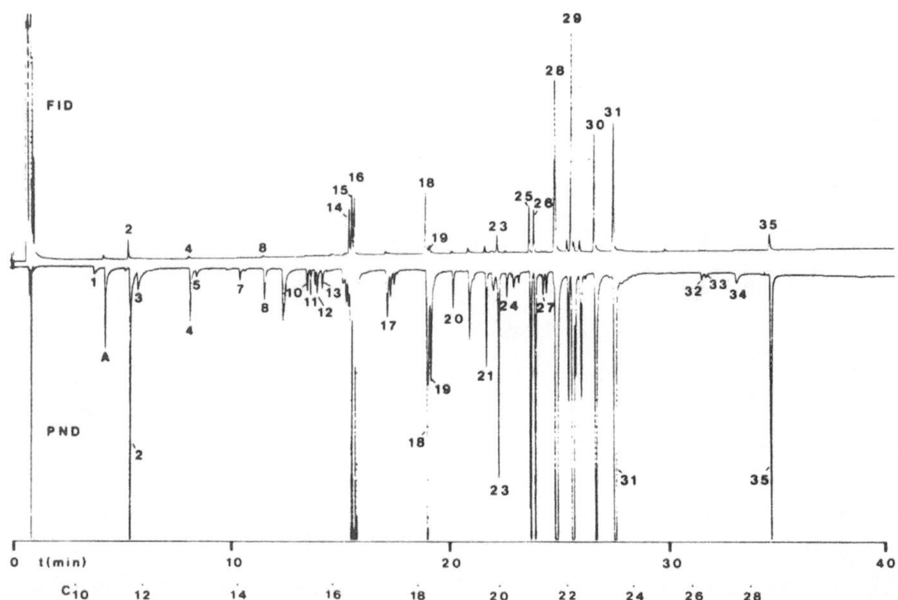

Fig. 3. GC chromatogram of *Datura innoxia* root extract obtained by FID and PND. *1* Hygrine; *2* tropine; *3* pseudotropine; *4* 3α-acetoxytropane; *5* 3β-acetoxytropane; *6* tyramine; *7* 3α-propionyloxytropane; *8* 3α-isobutyryloxytropane; *9* 3α-n-butyryloxytropane (synthetic); *10* 3α-2-methylbutyryloxytropane; *11* 3β-2-methylbutyryloxytropane; *12* N-methylpyrrolidinyl-hygrine A; *13* N-methylpyrrolidinyl-hygrine B; *14* 3α-tigloyloxytropane; *15* cuscohygrine; *16* 3β-tigloyloxytropane; *17* 3-hydroxy-6-(2-methylbutyryloxy)-tropane; *18* 3-tigloyloxy-6-hydroxytropane; *19* 3-hydroxy-6-tigloyloxytropane; *20* 3-tigloyloxy-6-acetoxytropane; *21* 3α-tigloyloxy-6β-propionyloxytropane; *22* meteloidine; *23* 3-tigloyloxy-6-isobutyryloxytropane; *24* apoatropine; *25* 3-tigloyloxy-6-(2-methylbutyryloxy)-tropane; *26* 3-tigloyloxy-6-methylbutyryloxytropane; *27* aposcopolamine; *28* hyoscyamine; *29* 3α-6β-ditigloyloxytropane; *30* scopolamine; *31* 3,6-ditigloyloxy-7-hydroxytropane; *32* 6-(2-methylbutyryloxy)-hyoscyamine; *33* 6-methylbutyryloxyhyoscyamine; *34* 6-tigloyloxyhyoscyamine; *35* unknown alkaloid. Positions of the coinjected hydrocarbons are marked below. *A* Artifact produced by hygrine. The separation was carried out with a DB-1 15 m × 0.25 mm i.d. fused-silica capillary column (Witte et al. 1987)

al. (1990) preferred to analyze the heptafluorobutyryl derivative of scopolamine after its extraction from rat plasma.

The analyses of the underivatized tropine alkaloids in the Solanaceae were usually carried out on medium length (25 m) apolar columns. Temperature programs ranged typically between 40 and 300 °C. Table 3 lists some examples of the systems used for the analysis of tropane alkaloids by GC.

The analysis of the pseudotropine alkaloids has been given considerable attention because of the abuse of cocaine. Cocaine diastereoisomers have been separated by GC using a packed column (Allen et al. 1981). Apolar columns separated the stereoisomers better than columns of medium

Table 3. Some GC systems used for the analysis of tropane alkaloids (for explanations to Footnote, see Table 1)

Column	GC conditions				Reference
Type, length (m) × i.d. (mm)	Injector	Temperature program	Detector	Carrier gas	
Tropine alkaloids					
DB-1[a], 15 × 0.25	250° s	70–300 6°/min	FID PND	He, 0.7 bar	Hartmann et al. (1986)
DB-1[a], 15 × 0.25	250° s	150–270° 6°/min	FID PND	He, 0.7 bar	Hartmann et al. (1986)
CP SIL 5cb[a], 10 × 0.3	250° l	40–210° 19°/min	FID	He, 2.5 ml/min	Ylinen et al. (1986)
Pseudotropine alkaloids					
OV-1[a], 12 × 0.2	230° s	150–250° 6°/min	MS	He, 83 kPa	Cooper and Allen (1984)

polarity. One of the diastereoisomers, pseudoallococaine, was less stable under GC conditions and the analysis was reported to be problematic. Other compounds of synthetic (Cooper and Allen 1984) and natural origin (LeBelle et al. 1988) could be separated without problems by capillary GC, but the quantitative analysis was carried out by HPLC-UV (254 nm) (LeBelle et al. 1988).

Moore et al. (1987) reported that it was not possible to chromatograph intact truxillines. These high molecular weight pseudotropine compounds degraded into the respective diphenylcyclobutane dicarboxylic acids and methyl ecgonidine, and this may explain in part why previous studies using GC systems did not report the presence of truxillines in coca leaf extracts. Later Lurie et al. (1990) reported that truxillines found in illicit samples of cocaine and in coca leaves could only be analyzed when the alkaloids were injected dissolved in CH_2Cl_2 – BSA (75:25, v/v). The applicability of GC for the analysis of illicit cocaine impurities and cocaine metabolites in animals and man was shown by several investigators (Jindal et al. 1979; Ambre et al. 1984; Jindal and Lutz 1986).

5 Quinoline Alkaloids

Quinoline alkaloids include quinine and quinidine which are widely used for treatment of respectively malaria and cardiac arrhythmias. Moreover, quinine is used as a bitter agent in beverages. Cinchonine and quinine were first shown to be separated on GC by Lloyd et al. (1960). A more recent publication describes the separation of the stereoisomers quinidine and quinine (Dagnino et al. 1991). Cinchonidine and cinchonine could not be separated under the same chromatographic conditions. Quinidine metabolites were analyzed successfully after silylation (Barrow et al. 1980; Leroyer et al. 1990).

6 Morphinan Alkaloids

The GC of morphinan alkaloids has forensic interest and has been reviewed recently by Hashimoto et al. (1988). Street samples of narcotics often contain adulterants such as procaine and lidocaine and diluents such as sugars. To be able to analyze the mixture of the different compounds simultaneously Comparini et al. (1983) silanized the samples. Good separation and peak shapes were obtained with this method for a variety of compounds.

For quantitative analysis, Bertol et al. (1989) found that a programmable temperature vaporizer inlet increased the accuracy of the analysis since it eliminated discrimination problems when analyzing compounds with a large range of molecular weights. With this method it was possible to increase the accuracy of the quantification of heroin, monacetylmorphine, acetylcodeine, papaverine, and narcotine commonly found in street samples of heroin.

Brenneisen and Borner (1985) identified and quantified thebaine, morphine, and codeine in different plant parts of *Papaver somniferum* and *P. bracteatum*. GC-MS of the acetate derivatives was used to confirm the identification obtained by HPLC. Perfluoroacylation was found to give, besides the usual derivatives, C-perfluoroacyl derivatives in the case of the presence of an enamine moiety in the ring D of the morphine-type molecule (Moore et al. 1984).

GC-MS was used to confirm the presence of thebaine (Kodaira et al. 1989) and codeine (Weitz et al. 1986) in mammalian brain.

7 Terpenoid Indole Alkaloids

Alkaloids of this class, like reserpine and vinblastine, play an important role in medicine. GC analysis of terpenoid indole alkaloids was shown to be possible by Lloyd et al. (1960). Later, this technique was further applied for various terpenoid indole alkaloids using packed columns (Verpoorte and Baerheim Svendsen 1984). More recently, capillary columns have been applied in the analysis of such alkaloids.

HPLC-UV is the preferred method for detection and quantification of terpenoid indole alkaloids but poor separation of some alkaloids (catharanthine, vindoline and ajmalicine) of *Catharanthus roseus* on isocratic systems have led Naaranlahti et al. (1989) and Ylinen et al. (1990) to develop a GC separation. Thin stationary phases were found to be essential to elute the compounds at reasonable temperatures and to achieve good separation and peak shapes.

The opportunity of obtaining further spectrometric data without laborious isolation procedures makes GC-MS analysis an attractive method for the analysis of terpenoid indole alkaloids. A list of 22 terpenoid indole alkaloids analyzed by GC was reported by Dagnino et al. (1991). Not all terpenoid indole alkaloids could be analyzed without previous derivatization procedures. An example of an unstable compound is catharanthine, which showed no analytically useful signal without previous derivatization (Ylinen et al. 1990).

Table 4 lists GC systems used for the analysis of terpenoid indole alkaloids and Fig. 4 illustrates a GC separation of a mixture of reference compounds.

8 Ergot Indole Alkaloids

In the past, contamination of wheat with ergot alkaloids occurred frequently and the consumption of contaminated grains has been the cause of thousands of deaths. With the introduction of suitable cleaning methods and agrochemicals, intoxications due to contaminated grains have become rare. Recent interest in the analysis of these compounds arose from the reoccurrence of intoxication cases related to the increasing demand for "alternative" products made from grains cultivated under natural conditions (Klug et al. 1988). Further interest arose from forensic laboratories in connection with the analysis of lysergic acid diethylamide (LSD).

The difficulty of obtaining a base line separation of LSD and its structural isomer lysergic acid methylpropylamide (LAMPA) using HPLC led

Table 4. Some GC systems used for the analysis of terpenoid indole alkaloids (for explanations to Footnote, see Table 1)

Column	GC conditions				Reference
Type, length (m) × i.d. (mm)	Injector	Temperature program (°C)	Detector	Carrier gas	
OV-1[a], 25 × 0.32	250° l	130° 2 min −280° 10°/min	FID 300°		Ylinen et al. (1990)
CP SIL 5cb[a], 10 × 0.22	220° s	100–175° 15°/min −230° 5°/min	FID 240°	N$_2$, 50 kPa	Dagnino et al. (1991)

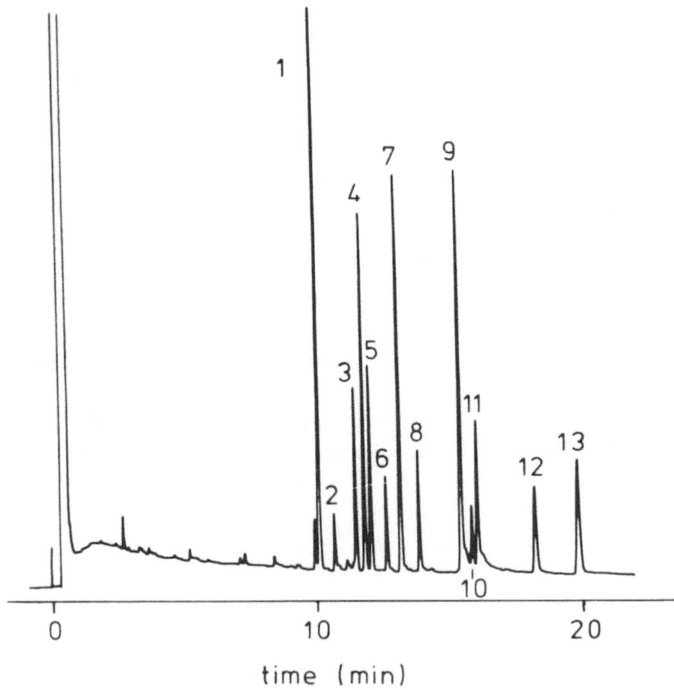

Fig. 4. Gas chromatogram of a mixture of terpenoid indole alkaloids: *1* Apparicine; *2* tubotaiwine; *3* tabersonine; *4* ibogamine; *5* voaphylline; *6* coronaridine; *7* aspidospermine; *8* pericyclivine; *9* perivine; *10* vobasine; *11* voacangine; *12* ibogaline; *13* ajmalicine. The GC conditions are described in Table 4. (Dagnino et al. 1991)

Japp et al. (1987) to develop a GC system to separate these compounds. A baseline separation was obtained but also under these conditions both compounds eluted closely together. To be able to rely on retention times it was necessary to check regularly the chromatographic resolution of the column before investigating unknowns. This was done by injecting a mixture of LSD and LAMPA or retention markers such as noscapine or etodroxizine, which elute between the two compounds of interest.

Ergot-peptide alkaloids were reported to be unstable under GC conditions but good separation of the degradation products of a mixture of these compounds can be obtained (Van Mansvelt et al. 1978). The mass spectra of the degradation products allowed the identification of the peptide part of the ergot alkaloids. Differences in the lysergic acid moiety could not be detected, since under the same conditions this part of the molecule was not recovered at all. Nevertheless, the combination of the information obtained through TLC, HPLC, and GC analysis should allow unequivocal identification of all ergot alkaloids. Klug et al. (1988) investigated ergot alkaloids in several food sources. HPLC was used for the identification and the quantification of ergot alkaloids. TLC and GC-NICI-MS were used to confirm the results obtained by HPLC. Feng et al. (1992) were able to control the degradation process occurring during the analysis by careful optimization of the chromatographic conditions. Only one degradation product of ergotamine was obtained and sensitive quantitative analysis in human plasma could be carried out.

9 Steroidal Alkaloids

Steroidal alkaloids are found in the Solanaceae. They occur in all parts of the plant and are generally glycosidically bound. Low levels of steroidal alkaloids can be detected in commercial varieties of potatoes and tomatoes. In breeding programs, wild species are sometimes used to introduce traits such as resistance to disease or cold. To avoid that, together with these traits, the ability to accumulate high levels of alkaloids is transmitted to hybrid progeny, wild species should be investigated with respect to their glyco-alkaloid composition (Van Gelder 1985).

Steroidal alkaloid glycosides have been investigated after derivatization by Herb et al. (1975) using a packed column. More recently, a capillary GC system was used for the separation and quantification of the underivatized aglycones (Van Gelder 1985). While developing the GC system, it was found that solasodine and tomatidine decomposed at high oven temperatures but that these compounds could be separated with less than 3% decomposition at 280 °C. Discrimination of the different compounds in a mixture was found to occur in the injector. The nonlinearity of the splitting process apparently increased at smaller sample volumes (Van Gelder et al. 1988), and so extra care was needed for quantitative analysis.

Steroidal alkaloids show very similar PND/FID response ratios, and this parameter can be used to distinguish them from other N-containing compounds in the chromatogram. This ratio, together with retention indices, can be used for the characterization of the steroidal alkaloids. For further identification, GC-MS could probably be restricted to compounds with PND/FID response ratios similar to the ones found for steroidal alkaloids and with retention indices which do not coincide with those of any known steroidal alkaloid (Van Gelder et al. 1988).

Medium length to long (25–50 m) apolar capillary columns were shown to be best suited for the analysis of these compounds (Van Gelder et al. 1988, 1989). Figure 5 shows a GC separation of C_{27} steroidal alkaloids. Since different biological activities are associated with structural differences in the sugar moieties, isolation and identification of the glycosides remain necessary to examine toxicity.

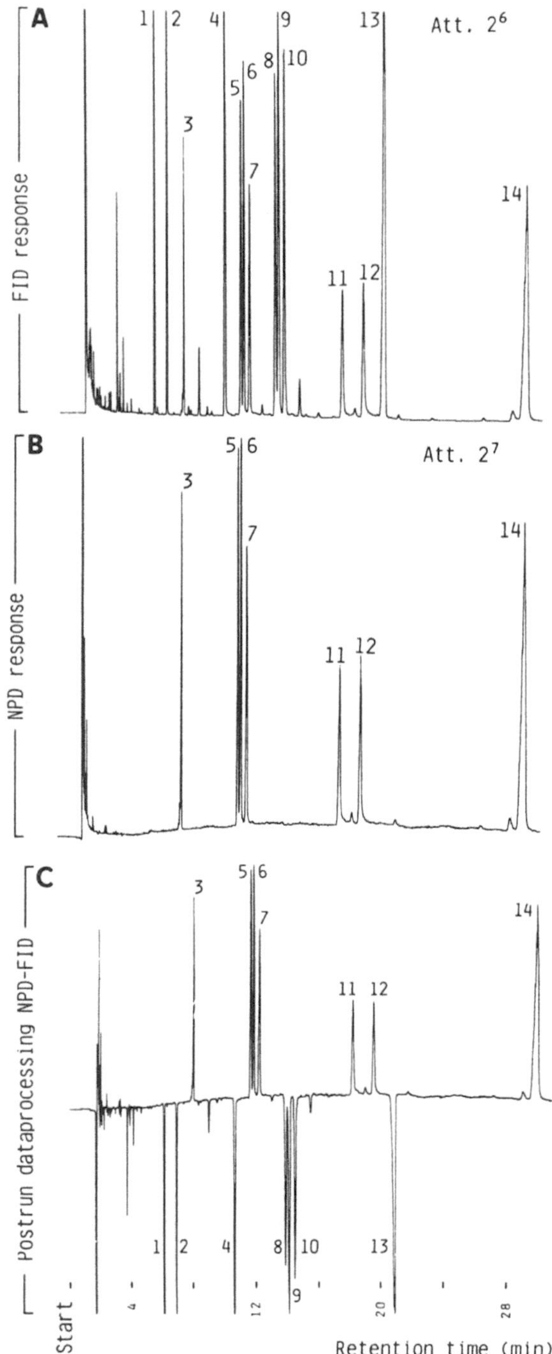

A
1 2 4 9 13 Att. 2⁶
8 10
5 6
3
7
11 12
14
FID response
B
5 6
3
7 Att. 2⁷
14
11 12
NPD response
C
5 6
3
7
14
11 12
1 2 4 8 10 13
9
Postrun dataprocessing NPD-FID
Start
4
12
20
28
Retention time (min)

10 Diterpenoid Alkaloids

Some *Delphinium* species are poisonous to cattle and cause losses in live-stock on mountain grazing ranges. Manners and Ralphs (1989) report the analysis of the alkaloid content of *D. occidentale* and *D. barbeyi* on a medium polarity column. Dictyocarpine and dictypocarpinine could not be separated under the conditions used and quantitative analysis of five alkaloids was shown to be possible. Figure 6 illustrates a GC separation of these compounds.

Table 5 gives a system used for the analysis of diterpenoid alkaloids.

11 Other Alkaloids

Lycopodium Alkaloids. Gerard and MacLean (1986) reported the analysis of *Lycopodium* species by GC-MS. Most of the alkaloids previously reported to occur in the species examined could also be found using the GC system. One of the exceptions was lycodiflexine, which might not be sufficiently volatile since its molecular weight (562) is well above that of the other components of the extract. Besides the alkaloids reported, other minor alkaloids could be detected by this method. The retention index of 15 alkaloids is included.

Vasicine and Related Alkaloids. Laakso et al. (1990) detected vasicine and related alkaloids in *Galega orientalis* Lam., a promising perennial forage legume for Finnish climatic conditions. Quantification was carried out by selective ion monitoring. The presence of galegine, a guanidine alkaloid, could not be confirmed by GC-MS, probably due to decomposition occurring during the analysis.

Polyhydroxylated Alkaloids. Polyhydroxy derivatives of pyrrolidine, piperidine, and indolizidine alkaloids have recently been isolated from plants and microorganisms. A number of these compounds have shown potent glucosidase inhibitory activity and have generated interest because of their ability to inhibit replication of retroviruses.

Fig. 5A–C. Gas chromatograms obtained simultaneously by FID (**A**) and PND (**B**) for a potato extract spiked with C_{27}-steroidal alkaloids, 5α-cholestane, sterols, and steroidal sapogenins. Attenuation: FID 2^6, PND 2^7. Post-analysis reprocessing of the raw data from **A** and **B** (subtraction of the FID trace from the PND trace) is shown in **C**. *1* Octacosane; *2* 5α-cholestane; *3* solanthrene; *4* cholesterol; *5* solanidine; *6* demissidine; *7* solasodiene; *8* stigmasterol; *9* diosgenin; *10* tigogenin; *11* solasodine; *12* tomatidine; *13* tetratriacontane; *14* jervine. Column: fused silica, 50 m × 0.22 mm i.d., CP Sil 5 cb, film thickness 0.12 um. Carrier gas: H_2, linear velocity 49 cm s^{-1}, oven temperature 270 °C (Van Gelder et al. 1988)

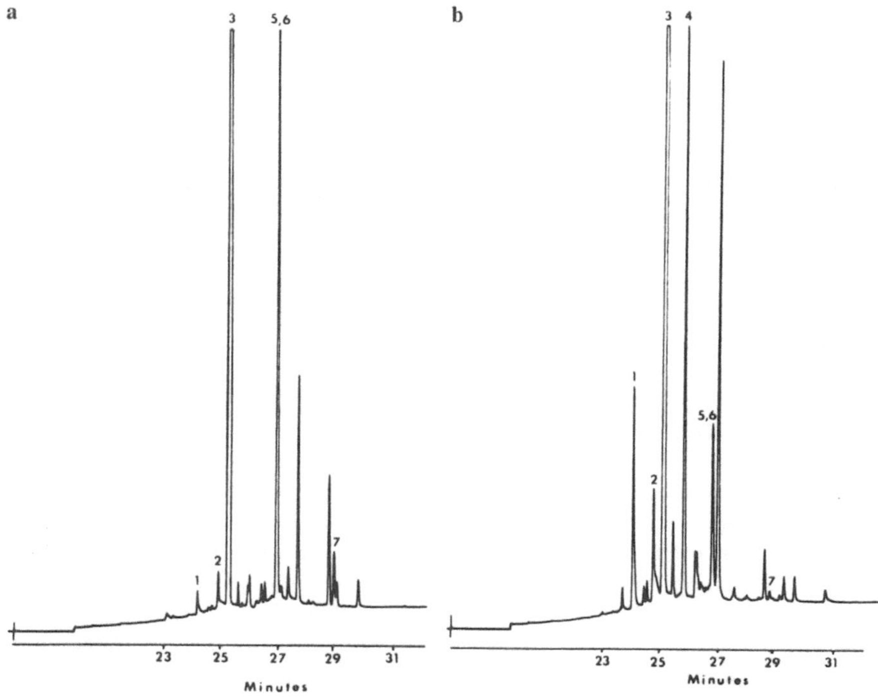

Fig. 6. a Capillary GC trace of total alkaloid extract of *Delphinium occidentale*. b Capillary GC trace of total alkaloid extract of *D. barbeyi*. Peaks: *1* Delpheline; *2* deltamine; *3* deltaline; *4* 14-acetyldictyocarpine; *5,6* dictyocarpine and dictyocarpinine; *7* delcosine. The GC conditions are described in Table 5. (Manners and Ralphs 1989)

Table 5. GC system used for the analysis of diterpenoid alkaloids

Column	GC conditions				Reference
Type, length (m) × i.d. (mm)	Injector	Temperature program (°C)	Detector	Carrier gas	
SE-30[a], 30 × 0.32		0.1 min 85° −175° 40°/min −300° 5°/min	FID 300°	He, 2.13 ml/ min	Manners and Pralphs (1989)

A gas chromatographic method for the separation of these compounds was reported by Nash et al. (1986). Trimethylsilyl derivatives of the poly-hydroxy alkaloids were analyzed in a packed column of medium polarity. More recently Molyneux et al. (1991) used a 30 m × 0.32 mm i.d. SE-30 capillary column to analyze the trimethylsilyl derivatives of this type of alkaloid.

12 Conclusion

In this chapter, the utility of GC in the analysis of alkaloids has been illustrated. Spectral information of unknown compounds from a mixture can be obtained by coupling a GC to a MS or to a FT-IR. For determination of known compounds in mixtures, other more common detectors can be used, such as FID, ECD (electron capture detection), or PND, for it is often sufficient to compare their retention time with the retention time of authentic samples. Furthermore, GC-MS also allows biosynthetic and metabolic studies using stable isotopes.

Accurate quantitative analysis requires detailed examination to obtain reproducible data. Errors which can arise due to characteristics of the GC system like different split ratios for different compounds, artifact formation, and degradation due to high temperatures should be minimized by careful optimization of the working conditions.

Not all compounds can be analyzed by GC. However, when it seems to be the only attractive method for the sensitive detection of the compounds of interest (for example in the case of compounds which do not contain a strong chromophore), increase in stability at higher temperatures or increase in volatility can be obtained through derivatization procedures or other modifications in the structure of the molecule (reduction of N-oxides, hydrolysis of glycoalkaloids).

No systematic studies to determine which column is best suited for alkaloid analysis have been reported, but from the methods described it becomes clear that thinly coated apolar columns are preferred for the analysis of underivatized alkaloids. Derivatized compounds have been analyzed frequently with columns of medium polarity.

The length of the columns used varies considerably (10 to 50 m). Anyway, it is advisable to test the stability of a compound under GC conditions with a short column. A longer column may be used later if the desired chromatographic resolution has not been achieved.

From the results described above it becomes clear that interesting information can be obtained by the analysis of alkaloids by GC and that especially the coupling to MS provides a wealth of information. Like HPLC, it is a useful method for the analysis of these compounds.

References

Allen AC, Cooper DA, Kiser WO, Cottrell RC (1981) The cocaine diastereoisomers. J Forensic Sci 26:12–26

Ambre J, Fischman M, Ruo T-I (1984) Urinary excretion of ecgoine methyl ester, a major metabolite of cocaine in humans. J Anal Toxicol 8:23–25

Ames MM, Powis G (1978) Determination of indicine N-oxide and indicine in plasma and urine by electron-capture gas-liquid chromatography. J Chromatogr 166:519–526

Balandrin MF, Kinghorn AD (1981) Tetrahydrorhombifoline, a further constituent of *Lupinus oscar-haughtii* and *L. truncatus*. J Nat Prod 44:495–497

Barrow SE, Taylor AA, Horning EC, Horning MG (1980) High-performance liquid chromatographic separation and isolation of quinidine and quinine metabolites in rat urine. J Chromatogr 181:219–226

Bertol E, Mari F, Di Milia MG (1989) Programmable temperature vaporizer applications in an high-resolution gas chromatographic method for the quantitation of impurities in illicit heroin. J Chromatogr 466:384–389

Bicchi C, D'Amato A, Cappelletti E (1985) Determination of pyrrolizidine alkaloids in *Senecio inaequidens* D.C. by capillary gas chromatography. J Chromatogr 349:23–29

Bicchi C, Caniato R, Tabacchi R, Tsoupras G (1989a) Capillary gas chromatography/positive and negative ion chemical ionization mass spectrometry on pyrrolizidine alkaloids of *Senecio inaequidens* using ammonia and hydroxyl ions as the reagent species. J Nat Prod 52:32–41

Bicchi C, Rubiolo P, Frattini C (1989b) Capillary gas chromatography-Fourier transform infra-red spectroscopy of pyrrolizidine alkaloids of *Senecio inaequidens* DC. J Chromatogr 473:161–170

Borstel K von, Witte L, Hartmann T (1989) Pyrrolizidine alkaloid patterns in populations of *Senecio vulgaris*, *S. vernalis* and their hybrids. Phytochemistry 28:1635–1638

Brenneisen R, Borner S (1985) Psychotrope Drogen. IV. Zur Morphinalkaloidführung von *Papaver somniferum* und *Papaver bracteatum*. Pharm Acta Helv 60:302–310

Chalmers AH, Culvenor CCJ, Smith LW (1965) Characterisation of pyrrolizidine alkaloids by gas, thin-layer and paper chromatography. J Chromatogr 20:270–277

Comparini IB, Centini F, Pariali A (1983) Simultaneous determination of narcotics, adulterants and diluents in street samples by means of gas chromatography with capillary columns. J Chromatogr 279:609–613

Cooper DA, Allen AC (1984) Synthetic cocaine impurities. J Forensic Sci 29:1045–1055

Dagnino D, Schripsema J, Peltenburg A, Verpoorte R, Teunis K (1991) Capillary gas chromatographic analysis of indole alkaloids: investigation of the indole alkaloids present in *Tabernaemontana divaricata* cell suspension culture. J Nat Prod 54:1558–1563

Deutsch J, Soncrant TT, Greig NH, Rapoport SI (1990) Electron-impact ionization detection of scopolamine by gas chromatography-mass spectrometry in rat plasma and brain. J Chromatogr 528:325–331

Dimenna GP, Krick TP, Segall HJ (1980) Rapid high-performance liquid chromatography isolation of monoesters, diesters and macrocyclic diester pyrrolizidine alkaloids from *Senecio jacobaea* an *Amsinckia intermedia*. J Chromatogr 192:474–478

Edgar JA, Culvenor CCJ (1974) Pyrrolizidine ester alkaloid in danaid butterflies. Nature 248:614–616

Feng N, Minder EI, Grampp T, Vonderschmitt DJ (1992) Identification and quantification of ergotamine in human plasma by gas chromatography-mass spectrometry. J Chromatogr 575:289–294

Gerard RV, MacLean DB (1986) GC/MS examination of four *Lycopodium* species for alkaloid content. Phytochemistry 25:1143–1150

Hartmann T, Toppel G (1987) Senecionine N-oxide, the primary product of pyrrolizidine alkaloid biosynthesis in root cultures of *Senecio vulgaris*. Phytochemistry 26:1639–1643

Hartmann T, Witte L, Oprach F, Toppel G (1986) Reinvestigation of the alkaloid composition of *Atropa belladonna* plants, root cultures and cell suspension cultures. Planta Med 52:390–395

Hashimoto T, Yamada Y (1983) Scopolamine production in suspension cultures and redifferentiated roots of *Hyoscyamus niger*. Planta Med 47:195–199

Hashimoto Y, Kawanishi K, Moriyasu M (1988) Forensic chemistry of alkaloids by chromatographic analysis, vol 32, chapter 1. In: Brossi A (ed) The Alkaloids. Academic Press, New York

Hatfield GM, Valdes LJJ, Keller WJ, Merrill WL, Jones VH (1977) An investigation of *Sophora secundiflora* seeds (Mescalbeans). Lloydia 40:374–383

Hatzold T, Elmadfa I, Gross R, Wink M, Hartmann T, Witte L (1983) Quinolizidine alkaloids in seeds of *Lupinus mutabilis*. J Agric Food Chem 31:934–938

Hendriks H, Balraadjsing W, Huizing HJ, Bruins AP (1987) Investigation into the presence of pyrrolizidine alkaloids in *Eupatorium cannabinum* by means of positive and negative ion chemical ionization GC-MS. Planta Med 53:456–461

Hendriks H, Huizing HJ, Bruins AP (1988) Ammonium positive-ion and hydroxide negative-ion chemical ionization gas chromatography-mass spectrometry for the identification of pyrrolizidine alkaloids in *Eupatorium rotundifolium* L. var. *ovatum*. J Chromatogr 428:352–356

Herb SF, Fitzpatrick ThJ, Osman SF (1975) Separation of glycoalkaloids by gas chromatography. J Agric Food Chem 23:520–523

Hesse M (1974) Progress in mass spectrometry, vol 1, parts 1 and 2. Mass spectrometry of indole alkaloids. Verlag Chemie, Weinheim

Hesse M, Bernhard HO (1975) Progress in mass spectrometry, vol 3. Mass spectrometry of alkaloids. Verlag Chemie, Weinheim

Huizing HJ, De Boer F, Malingre ThM (1981) Preparative ion-pair high-performance liquid chromatography and gas chromatography of pyrrolizidine alkaloids from comfrey. J Chromatogr 214:257–262

Japp M, Gill R, Osselton MD (1987) The separation of lysergide (LSD) from related ergot alkaloids and its identification in forensic science casework samples. J Forensic Sci 32:933–940

Jindal SP, Lutz T (1986) Ion cluster techniques in drug metabolism: use of labelled and unlabelled cocaine to facilitate metabolite identification. J Anal Toxicol 10:150–155

Jindal SP, Lutz T, Vestergaard P (1979) Gas-liquid chromatographic-mass spectrometric determination of lidocaine in an illicit sample of cocaine. J Chromatogr 179:357–360

Keller WJ, Zelenski SG (1978) Alkaloids from *Lupinus argenteus* var. *stenophyllus*. J Pharm Sci 67:430–431

Keller WJ, Meyer BN, McLaughlin JL (1983) (−)-alpha-Isosparteine from *Lupinus argenteus* var. *stenophyllus*. J Pharm Sci 72:563–564

Kinghorn AD, Selim MA, Smolenski SJ (1980) Alkaloid distribution in some new world *Lupinus* species. Phytochemistry 19:1705–1710

Kinghorn AD, Balandrin MF, Lin L-J (1982) Alkaloid distribution in some species of the papilionaceous tribes Sophoreae, Dalbergieae, Loteae, Brongniartieae and Bossiaeeae. Phytochemistry 21:2269–2275

Klug C, Baltes W, Kroenert W, Weber R (1988) Methode zur Bestimmung von Mutterkornalkaloiden in Lebensmitteln. Z Lebensm Unters Forsch 186:108–113

Kodaira H, Lisek CA, Jardine I, Arimura A, Spector S (1989) Identification of the convulsant opiate thebaine in mammalian brain. Proc Natl Acad Sci USA 86:716–719

Laakso I, Virkajaervi P, Airaksinen H, Varis E (1990) Determination of vasicine and related alkaloids by gas chromatography-mass spectrometry. J Chromatogr 505:424–428

LeBelle MJ, Callahan SA, Latham DJ, Lauriault G (1988) Identification and determination of norcocaine in illicit cocaine and coca leaves by gas chromatography-mass spectrometry and high-performance liquid chromatography. Analyst 113:1213–1215

Leroyer R, Varoquaux O, Advenier C, Pays M (1990) Quinidine oxidative metabolism. Identification and biosynthesis of quinidine 10,11-dihydrodiol stereoisomers. Biomed Chromatogr 4:61–64

Lloyd HA, Fales HM, Highet PF, VandenHeuvel WJA, Wildman WC (1960) Separation of alkaloids by gas chromatography. J Am Chem Soc 82:3791

Luethy J, Zweifel U, Schmid P, Schlatter Ch (1983) Pyrrolizidin-Alkaloide in *Petasites hybridus* L. und *P. albus* L. Pharm Acta Helv 58:98–100

Lurie IS, Moore JM, Kram TC, Cooper DA (1990) Isolation, identification and separation of isomeric truxillines in illicit cocaine. J Chromatogr 504:391–401

Manners GD, Ralphs MH (1989) Capillary gas chromatography of *Delphinium* diterpenoid alkaloids. J Chromatogr 466:427–432

Mattocks AR, Jukes R (1990) Recovery of the pyrrolic nucleus of pyrrolizidine alkaloid metabolites from sulphur conjugates in tissues and body fluids. Chem-Biol Interact 75:225–239

Molyneux RJ, Pan YT, Tropea JE, Benson M, Kaushal GP, Elbein AD (1991) 6,7-di*epi*castanospermine, a tetrahydroxyindolizidine alkaloid inhibitor of amyloglucosidase. Biochemistry 30:9981–9987

Moore JM, Allen AC, Cooper DA (1984) Determination of manufacturing impurities in heroin by capillary gas chromatography with electron capture detection after derivatization with heptafluorobutyric anhydride. Anal Chem 56:642–646

Moore JM, Cooper DA, Lurie IS, Kram TC, Carr S, Harper C, Yeh J (1987) Capillary gas chromatographic-electron capture detection of coca-leaf-related impurities in illicit cocaine: 2,4-diphenylcyclobutane-1,3-dicarboxylic acids, 1,4-diphenylcyclobutane-2,3-dicarboxylic acids and their alkaloidal precursors, the truxillines. J Chromatogr 410:297–318

Muehlbauer P, Witte L, Wink M (1988) New ester alkaloids from lupines (genus *Lupinus*). Planta Med 54:237–239

Naaranlahti T, Lapinjoki SP, Huhtikangas A, Toivonen L, Kurten U, Kauppinen V, Lounasmaa M (1989) Mass spectral evidence of the occurrence of vindoline in heterotrophic cultures of *Catharanthus roseus* cells. Planta Med 55:155–157

Nash RJ, Goldstein WS, Evans SV, Fellows LE (1986) Gas chromatographic method for separation of nine polyhydroxy alkaloids. J Chromatogr 366:431–434

Parr AJ, Payne J, Eagles J, Chapman BT, Robins RJ, Rhodes MJC (1990) Variation in tropane alkaloid accumulation within the Solanaceae and strategies for its exploitation. Phytochemistry 29:2545–2550

Pieters LA, Hartmann T, Janssens J, Vlietinck AJ (1989) Comparison of capillary gas chromatography with ^{1}H an ^{13}C nuclear magnetic resonance spectroscopy for the quantitation of pyrrolizidine alkaloids from *Senecio vernalis*. J Chromatogr 462:387–391

Popl M, Fähnrich J, Tatar V (1990) Chromatographic analysis of alkaloids. Marcel Dekker, New York

Priddis CR (1983) Capillary gas chromatography of lupin alkaloids. J Chromatogr 261:95–101

Roeder E, Neuberger V (1988) Pyrrolizidinalkaloide in *Symphytum*-Arten. Ein Beitrag zum qualitativen und quantitativen Nachweis. Dtsch Apoth Ztg 128:1991–1994

Simola LK, Martinsen A, Huhtikangas A, Jokela R, Lounasmaa M (1989) Feeding experiments with precursors of tropane alkaloids using suspension cultures of *Atropa belladonna*. Acta Chem Scand 43:702–705

Stelljes MA, Kelley RB, Molyneux RJ, Seiber JN (1991) GC-MS determination of pyrrolizidine alkaloids in four *Senecio* species. J Nat Prod 54:759–773

Stengl P, Wiedenfeld H, Roeder E (1982) Lebertoxische Pyrrolizidinalkaloide in *Symphytum*-Präparaten. Ein Beitrag zum sicheren und schnellen Bestimmung der Alkaloide in Handelsdrogen und pharmazeutischen Zubereitungen. Dtsch Apoth Ztg 122:851–855

Strack D, Becher A, Brall S, Witte L (1991) Quinolizidine alkaloids and the enzymatic syntheses of their cinnamic and hydroxycinnamic acid ester in *Lupinus angustifolius* and *L. luteus*. Phytochemistry 30:1493–1498

Van Gelder WMJ (1985) Determination of the total C_{27}-steroidal alkaloid composition of *Solanum* species by high-resolution gas chromatography. J Chromatogr 331:285–293

Van Gelder WMJ, Jonker HH, Huizing HJ, Scheffer JJC (1988) Capillary gas chromatograpy of steroidal alkaloids from Solanaceae. Retention indices and simultaneous flame ionization nitrogen-specific detection. J Chromatogr 442:133–145

Van Gelder WMJ, Tuinstra LGMTh, Van der Greef J, Scheffer JJC (1989) Characterization of novel steroidal alkaloids from tubers of *Solanum* species by combined gas chromatography-mass spectrometry. Implications for potato breeding. J Chromatogr 482:13–22

Van Mansvelt FJW, Greving JE, De Zeeuw RA (1978) Identification of ergot-peptide alkaloids, based on gas-liquid chromatography of the peptide moiety. J Chromatogr 151:113–120

Verpoorte R, Baerheim Svendsen A (1984) Chromatography of alkaloids. Part B. Gasliquid chromatography and high-performance liquid chromatography. J Chromatogr Libr, vol 23B. Elsevier, Amsterdam

Weitz CJ, Lowney LI, Faull KF, Feistner G, Goldstein A (1986) Morphine and codeine from mammalian brain. Proc Natl Acad Sci USA 83:9784–9788

Wink M, Hartmann T (1988) Production of quinolizidine alkaloids by photomixotrophic cell suspension cultures: biochemical and biogenetic aspects. Planta Med 40:149–155

Wink M, Witte L, Schiebel HM, Hartmann T (1980) Alkaloid pattern of cell suspension cultures and differentiated plants of *Lupinus polyphyllus*. Planta Med 38:238–245

Wink M, Witte L, Hartmann T (1981a) Quinolizidine alkaloid composition of plants and of photomixotrophic cell suspension cultures of *Sarothamnus scoparius* and *Orobanche rapumgenistae*. Planta Med 43:342–352

Wink M, Hartmann T, Witte L, Schiebel HM (1981b) The alkaloid patterns of cell suspension cultures and differentiated plants of *Baptisia australis* and their biogenetic implications. J Nat Prod 44:14–20

Wink M, Schiebel HM, Witte L, Hartmann T (1982) Quinolizidine alkaloids from plants and their cell suspension cultures. Planta Med 44:15–20

Wink M, Witte L, Hartmann T, Theuring C, Volz V (1983) Accumulation of quinolizidine alkaloids in plants and cell suspension cultures: genera *Lupinus*, *Cytisus*, *Baptisia*, *Genista*, *Laburnum* and *Sophora*. Planta Med 48:253–257

Winter CK, Segall HJ, Jones AD (1988) Determination of pyrrolizidine alkaloid metabolites from mouse liver microsomes using Tandem mass spectrometry and gas chromatography/mass spectrometry. Biomed Environ Mass Spectrom 15:265–273

Witte L, Mueller K, Arfmann H-A (1987) Investigation of the alkaloid pattern of *Datura innoxia* plants by capillary gas-liquid-chromatography-mass spectrometry. Planta Med 53:192–197

Witte L, Ernst L, Adam H, Hartmann T (1992) Chemotypes of two pyrrolizidine alkaloid-containing *Senecio* species. Phytochemistry 31:559–565

Yamada Y, Hashimoto T (1982) Production of tropane alkaloids in cultured cells of *Hyoscyamus niger*. Plant Cell Rep 1:101–103

Ylinen M, Naaranlahti T, Lapinjoki S, Huhtikangas A, Salonen M-L, Simola LK, Lounasmaa M (1986) Tropane alkaloids from *Atropa belladonna*. Part I. Capillary gas chromatographic analysis. Planta Med 52:85–87

Ylinen M, Suhonen P, Naaranlahti T, Lapinjoki S, Huhtikangas A (1990) Gas chromatographic-mass spectrometric analysis of major indole alkaloids of *Catharanthus roseus*. J Chromatogr 505:429–434

Alkaloid Analysis in Flue-Cured Tobacco

C.A. WILKINSON and W.W. WEEKS

1 Introduction

Tobacco (*Nicotiana tabacum* L.) is the most widely grown commercial non-food plant, with world production at 2.4 million ha in 1991 (USDA-FAS 1991). It is a New World crop that has alternately been valued for its medicinal properties and condemned as being a health hazard. Although there are 64 *Nicotiana* species currently known (Sisson and Severson 1990), all commercial tobacco belongs to *Nicotiana tabacum* L. Commercial tobaccos are cultivated for their cured leaf and are classified according to their field production and method of leaf curing (Akehurst 1981). Flue-cured tobacco is one of several classes, it occupies predominate hectarages in the world, and is one of the principal classes used in the manufacture of cigarettes.

Angiosperms, flowering plants, have been the major source of alkaloids, and the Solanaceae, of which tobacco is a member, are one of the most important alkaloid-containing families. True alkaloids contain nitrogen in a heterocyclic ring, are toxic basic compounds derived from amino acids, occur in plants as salts of an organic acid, and show a wide range of physiological activities. More than 20 different alkaloids have been identified in tobacco, of which nicotine, a pyridine alkaloid, is the predominate one in flue-cured tobacco (Enzell et al. 1977; Bush 1981). It is the only known physioactive compound in tobacco and makes the plant unique. The stimulatory effect of nicotine is responsible for the development and use of tobacco around the world. Nicotine is the most extensively studied and most important alkaloid in leaves, and is considered a primary alkaloid in tobacco. A number of alkaloids occur in flue-cured tobacco in minute quantities, referred to as secondary or minor alkaloids (Tso 1990), and are a deterrent to smoking quality (Mosley and Rayburn 1957). Minor tobacco alkaloids include anabasine, anatabine, myosmine, and nornicotine.

The flue-cured leaf is the only portion of the plant used in manufacturing tobacco products. Distribution of alkaloids in flue-cured tobacco leaves varies by stalk position and within individual leaves regardless of stalk position. Alkaloids readily form organic salts with dicarboxylic acids and amino acids; alkaloids also form complexes with proteins, polyphenols, carbonyls, and iron to produce color in tobacco leaves (Dawson 1945). Alkaloid complexes in the cured leaf are water-soluble; however, aqueous

Modern Methods of Plant Analysis, Volume 15
Alkaloids (ed. by Linskens/Jackson)
© Springer-Verlag Berlin Heidelberg 1994

extracts are made basic to free the alkaloid salt as a free compound during extraction. Aqueous mineral acids used during extraction of alkaloids from plant material also form salts with alkaloids. Increasing the alkalinity of these aqueous acid solutions results in alkaloids becoming free compounds. Alkaloids can be removed from aqueous basic solution with organic solvents such as chloroform, ether, methylene chloride, and other organics that are immiscible with water.

Flavor and aroma of the cured leaf and smoke are directly related to leaf chemistry. Tobacco quality and usability are influenced by total alkaloid content. Determination of alkaloid levels is important to all aspects of tobacco production and research. Analyses of alkaloids are employed by geneticists to determine the performance of new germplasm, by agronomists to ascertain the influence of new cultural practices, by agricultural engineers to evaluate different curing regimes, and by manufacturers of tobacco products to control blending operations.

This chapter is not intended to include a detailed description of all possible methods of alkaloid analysis in tobacco nor provide a comprehensive review of the literature. Certain methods of historical or practical significance have been highlighted with the most pertinent papers cited for each method. The methods of alkaloid analysis discussed are applicable to all classes of tobacco.

2 Factors Affecting Alkaloid Concentration

The physical and chemical properties of tobacco leaves are influenced by genetics, agricultural practices, soil type, nutrients, weather conditions, plant disease, stalk position, harvesting, and curing procedures (Hawks and Collins 1983). The alkaloids produced and the inherent production capacity of each is determined by the genetic makeup of the plant (Mann et al. 1975). Weather, soil type, fertilization, and numerous management variables associated with the culture and curing processes influence the actual quantity of the individual alkaloids to varying degrees (Chaplin and Miner 1980; Campbell et al. 1982a,b). A change in any of these factors can markedly alter leaf composition and thus affect smoking quality. Total alkaloid concentration in the leaves ranges between 0.2 and 8% (Tso 1990).

The rapid and wide-ranging fluctuations in environmental conditions such as temperature, solar radiation, humidity, and soil moisture can significantly influence the chemical composition of the leaf. An increase in day/ night temperature combinations is generally correlated with total alkaloid levels, presumably as a result of increased synthesis and translocation of nicotine in response to increased root metabolism and transpiration (Long and Woltz 1977). Total alkaloid levels were highest in the lower leaves at high temperatures but highest in the upper leaves at low temperatures. Increased day lengths have been reported to increase (Tso et al. 1970) and decrease (Long and Woltz 1977) alkaloid concentrations. Accumulation of alkaloids is generally greater when humidity is high during the day. It has been suggested that high photosynthetic capacity in response to high humidity also promotes increased root growth, thereby increasing alkaloid synthesis. Severe cases of frost, hail, and air pollution stress have also been reported to significantly impact the chemical quality of the cured leaf (Mulchi 1985).

Alkaloid content of roots and leaves at flowering increased with moisture deficiency (van Bevel 1953; Weybrew et al. 1983). Drought stress tends to slow growth rate and reduce leaf area, which results in increased nicotine content. Excess rainfall increases growth rate and leaf size and may reduce the available nitrogen by leaching, resulting in decreased nicotine content.

3 Methods of Analysis of Tobacco Alkaloids

The chemistry of pyridine alkaloids has resulted in many different methods for determining flue-cured tobacco alkaloids qualitatively and quantitatively. Some of these older methods are rather laborious and time-consuming, but may be applicable in the absence of modern techniques. The major alkaloid of interest in flue-cured tobacco is nicotine, which comprises 95% of the total alkaloids in domestic cultivars; therefore, it is only necessary to determine total alkaloids as nicotine in most cases. Most laboratories currently use methods for alkaloid analysis which are rapid, sensitive, and provide more dependable data. The introduction of preparative chromatographic techniques and sophisticated spectroscopic instrumentation has led to a dramatic increase in the number of known alkaloids.

3.1 Steam Distillation

Pyridine alkaloids are high boiling compounds which are stable to high temperatures. Nicotine, the most volatile of the pyridine alkaloids, is readily steam distilled from strong basic solutions; however, secondary alkaloids are less volatile than nicotine and are not easily steam distilled quantitatively from basic aqueous solutions. Since nicotine is the predominant pyridine alkaloid present in most tobaccos, this analysis suffices for total alkaloids because secondary alkaloid concentrations are minimal. Tobacco alkaloids removed from basic aqueous solutions by steam distillation are quantified with a spectrophotometer.

A complete steam distillation method, which modified earlier distillation apparatus, was semi-automated by Griffith and Jeffrey (1948) and Griffith

(1957). The apparatus consists of heaters, the still, a connecting tube, a condenser, U-tube receiver, and a collection flask. A tobacco sample ground to pass a 60-mesh screen is introduced into 5 ml of salt saturated 30% base (NaOH) in the still. Steam is introduced into the sample and the distillate is collected over 10 ml of 3 N H_2SO_4 until the volume reaches 250 ml. The acidic distillate is read at three wavelengths: 236, 259, and 282 nm. Percent total alkaloids calculated as nicotine are determined using the following equation: $[D_{259} - (D_{236} + D_{282}/2) \times 0.780]/g$ sample weight.

Analysis is most efficient with a Griffith Still which consists of a series of six stills connected together with a master electrical control. An experienced operator can run 40 samples/h (Griffith and Jeffery 1948; Griffith 1957). The Griffith method is an AOAC (Association of Official Analytical Chemists)-approved method for determination of total alkaloids in tobacco (Williams 1984). Steam distillation is not the method of choice for determining secondary alkaloids which include nornicotine, anabasine, anatabine, and myosmine. In cases where individual alkaloids in a sample must be analyzed, steam distillation is not dependable. However, in cases where total alkaloids determined as nicotine will suffice, steam distillation is adequate and the use of more sensitive equipment is not necessary.

3.2 Autoanalyzer

An autoanalyzer method was developed to determine total alkaloids using nicotine as a standard (Sadler et al. 1960; Harvey et al. 1969) to accommodate the large volume of samples routinely determined each year by tobacco companies, leaf dealers, and tobacco researchers. This method is rapid and has the added advantage of simultaneously determining nicotine alkaloids and reducing sugars in flue-cured tobacco. The autoanalyzer system consists of a rotating sample table, sampling tubes, a partition pump, mixing coils, a 6.096 m delay coil, colorimeter, and a strip chart recorder.

The reagents and standards used include an extraction solution [10% glacial acetic acid and 4% methanol (v/v) diluted to volume in deionized water], cyanogen bromide solution, sodium hydroxide solution (36 g reagent grade NaOH/l), buffered aniline solution, and nicotine standards (Harvey et al. 1969). For the cyanogen bromide solution, 20 g of reagent grade cyanogen bromide (CNBr) and 0.5 ml of BRIJ-35 surfactant are dissolved in deionized distilled water and brought to volume in a 1-l volumetric flask. The buffered aniline solution is made by adding 11.24 g of reagent grade sodium phosphate dibasic (Na_2HPO_4), 8.2 g reagent grade citric acid monohydrate ($C_6H_8O_7 \cdot H_2O$), and 3 ml of reagent grade aniline to 900 ml of deionized distilled water in a 1-l volumetric flask. Bring to volume with deionized distilled water after these are dissolved. Nicotine standards for calibration are prepared by adding 0.01, 0.03, 0.05, 0.06, 0.08, and 0.10 mg nicotine/ml in extracting solution, treated with charcoal (0.5 g/100 ml), and filtered through Whatman #1 filter paper to remove the charcoal from the standards. The standards should bracket percent nicotine alkaloids expected from unknown samples. A known tobacco sample may be added after each set of standards as an estimate of reproducibility.

A 0.200-g ground sample of tobacco is weighed into a 250-ml Erlenmeyer flask with a ground glass stopper. Also added to the flask is 0.5 g of activated charcoal (Darco G-60 phosphate-free) to remove interfering color, and 100 ml of extraction solution. The samples are extracted for 5 min on a fast action shaker and then filtered through Whatman #1 filter paper.

The sampler operates off a rotating cam that drives the sampler and rotates the sample table from one sample to the next. The cam is constructed so that the sample is picked up for 0.5 min at the rate of 0.3 ml/min and the extracting solution (wash solution) is picked up for 1.0 min at the rate of 0.6 ml/min. The reagents are picked up constantly with a partitioning pump: sodium hydroxide (0.60 ml/min), air (0.80 ml/min), cyanogen bromide (1.6 ml/min), buffered aniline (2.9 ml/min), and air (1.6 ml/min). Air is used to move reagents and samples through the manifold, where mixing occurs, with a minimum of spreading. The acidic sample and NaOH reagent are thoroughly mixed in a mixing coil to neutralize the acidic sample. Cyanogen bromide and aniline are mixed in a second mixing coil. As the cyanogen bromide and aniline mixture leave the mixing coil, 0.8 ml of the neutralized sample is reintroduced (0.4 ml is discarded) into cyanogen bromide and aniline for mixing before the total mixture enters the delay coil. Full color develops from the reaction of alkaloids, cyanogen bromide, and aniline in the time delay coil before entering the flow cell. The reaction mixture passes through a 15-mm flow cell in the colorimeter where absorbance is read at 460 nm. The excess solution that does not pass through the flow cell is discarded as waste. The sample rate for the autoanalyzer is 40 samples/h with a 2:1 wash cycle between samples. The extracting solution is used for the wash. A diagram of the autoanalyzer manifold (Fig. 1) demonstrates the mechanics for this system.

Only a fraction of the total solution is needed for color development. The baseline is established with extraction solution flowing through the colorimeter at the start of the analyses. Nicotine standards are run in an ascending concentration prior to the tobacco samples and in descending concentration following the tobacco samples. Standards should be run following two sample trays (80 samples). Samples spiked with nicotine or a check of know concentration are placed midway through each tray to check the system. Duplicate samples or duplicate trays can be run to determine the reproducibility of the system.

Current autoanalyzers are linear over the range of analytical standards used. Absorbance values are read directly from the chart. The method of least squares is used to calculate the standard curve where the concentration of nicotine standards is X and absorbance of nicotine (from the chart) is Y. The equation of a straight line ($Y = mX + b$) is used to calculate the

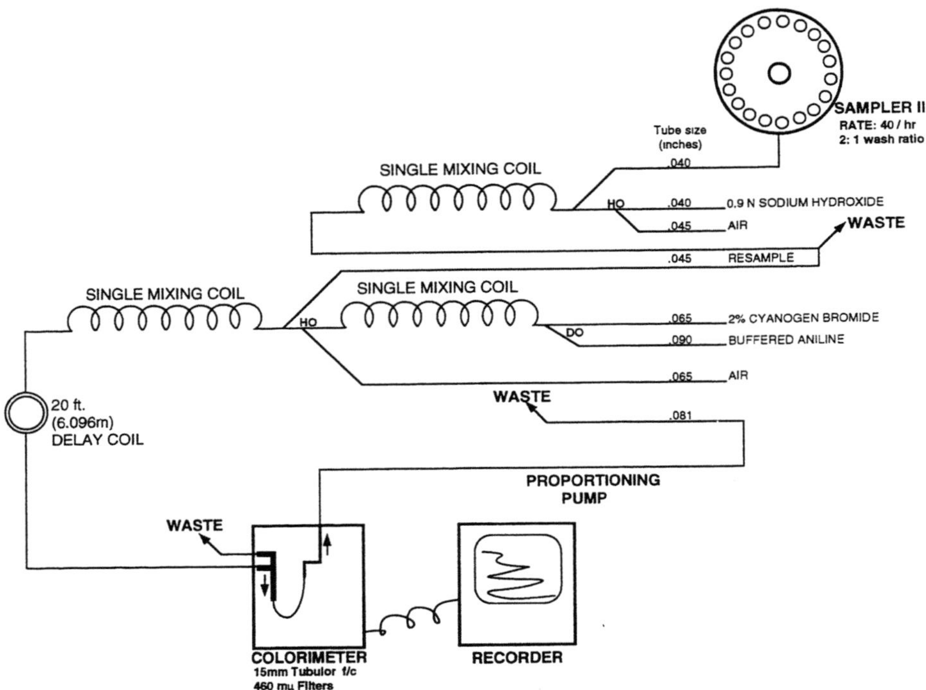

Fig. 1. Diagram of autoanalyzer manifold for analysis of nicotine in tobacco extracts

concentration of each sample where Y = absorbance value from the chart, m = slope of the line, b = the intercept, and X (calculated) = the concentration in 1 ml of the sample. The total alkaloid concentration is determined by the following equation:

$$\% \text{ total alkaloids} = [X/\text{weight of sample in mg}] \times 100.$$

3.3 Chromatographic Analysis of Tobacco Alkaloids

Tobacco breeders have found it necessary to use wild species of *Nicotiana* in breeding programs to introduce disease and insect resistance into new flue-cured varieties. Such practices often result in tobaccos that produce undesirably high levels of nornicotine, anabasine, and anatabine. Colorimetric analyses used to determine nornicotine do not give accurate results if anabasine and anatabine are produced in high amounts, since the methods used for determining nornicotine do not account for anabasine, anatabine, and myosmine quantitatively. Chromatographic procedures were developed that allowed tobacco breeders to separate complex mixtures of alkaloids into individual alkaloids.

3.3.1 Paper Chromatography

Paper chromatography was the first useful chromatographic method developed and used that gave satisfactory separations of pyridine alkaloids from tobacco (Jeffrey and Tso 1955; Jeffery and Eoff 1955). Paper chromatography is slow, laborious, and is not as sensitive as some of the more modern methods, but can still be used by breeders today when absolutely necessary. Either quantitative or qualitative determinations are possible with paper chromatography and both green and cured tobacco may be used.

Analyses of cured tobacco requires at least 2.0 to 2.5 g of ground tobacco extracted overnight with 25 ml of 1:1 (v/v) acetone and water. The sample is mixed thoroughly, allowed to set overnight with frequent shaking, and filtered into test tubes for spotting. Preparation of green tobacco samples requires 50 to 100 g of fresh tissue. In order to obtain an efficient extraction in green tissue, particularly if the sample is to be quantified, the water in the green tobacco must be taken into account. Acetone is added to the green leaf sample (equivalent to water in the tissue) which is then ground in a Waring blender, filtered into a test tube, and spotted on chromatography paper. Large rectangular sheets of dried Whatman #1 paper that have been pretreated with 0.1 M (pH 6) phosphate buffer (Na_2HPO_4) are spotted with 10 μl of each extract in a line 3.8 cm from the edge of the paper. Ten μl of nicotine, nornicotine, anabasine, anatabine, and myosmine are spotted opposite the plant samples as standards.

Large paper chromatographic jars are equilibrated to tank saturation, 4 to 16 h, with developing solvent (100 ml of *tert*-amyl alcohol, 20 ml of water, and 0.3 g of ethyl *p*-aminobenzoate). The chromatographic sheets are developed in an ascending manner overnight and air-dried in a chromatogram cabinet or hood. Dry sheets are sprayed with 0.2 M (pH 5.6) acetate buffer and placed in a closed chromatographic tank containing cyanogen bromide crystals. The chromatograms are removed from the cyanogen bromide tanks after 5 min. Cyanogen bromide is allowed to evaporate from the paper in a hood. Color of the spots developed is compared with the standard, and the concentration of each spot is approximated with the concentration of the standard. Concentration of each tobacco sample is determined by weighing the extracts (1 g/ml) or determining the total volume of the extract multiplied by the amount of each alkaloid determined from the volume (10 μl) spotted on the paper.

Further quantification can be obtained from paper chromatography by streaking paper with 1 ml or more of tobacco extract. Drying the paper while streaking will avoid migration of the band while impregnating the paper. The bands are chromatogrammed and developed as previously described. Alkaloid bands are located on the paper with UV light and marked. Bands are cut out, the alkaloid band is removed from the paper with chloroform, and the solution is read with a spectrophotometer after establishing the absorbance maximum. Absorbance maximum and a standard curve must be determined for nicotine and nornicotine, which will suffice for all of the secondary alkaloids (Stephens and Weybrew 1959).

3.3.2 Paper Chromatography – Ultraviolet Spectrophotometer

In the development of disease resistant varieties, secondary alkaloids are often a concern for flue-cured tobacco breeders. A rapid color test developed by Glock and Wright (1963) is used to quantify nornicotine in flue-cured tobacco. The method involves color development from the reaction of nornicotine and 1,3-indanedione. This test is sensitive enough to detect microgram levels of nornicotine. The method is used quantitatively with a spectrophotometer because color development is linear from 5.0 to $100\,\mu g/ml$ of nornicotine reacted with 1,3-indanedione.

A 100- to 1000-mg sample of ground air-dried flue-cured tobacco is weighed into a 125-ml Erlenmeyer flask. Ten ml of $2\,N$ NaOH and 50 ml of chloroform are added to the flask. The flask is shaken vigorously for 30 min and then 2 g of analytical grade celite are added to absorb water. The extraction process is completed by filtering the tobacco extract through two layers of Whatman #40 paper.

The assay consists of adding from 1 to 3 ml of tobacco extract (do not exceed $100\,\mu g$) to a test tube, along with the p-hydroxybenzoic acid reagent (do not exceed 6 ml), and 1 ml of indanedione solution. The p-hydroxybenzoic acid reagent is prepared by adding 2 g of p-hydroxybenzoic acid to a solution of 80% acetone and 20% absolute ethanol (v/v). High purity acetone is required because of the sensitivity of the indanedione reaction to impurities. The indanedione solution consists of a 0.2% solution of purified 1,3-indanedione dissolved in chloroform (w/v) and then stored in an amber bottle. The solutions in the test tube are thoroughly mixed using a Vortex mixer, and the color is allowed to develop for 1 h. Absorbance is determined at 550 nm after correcting for reagent blanks and tobacco extract blanks. A standard curve is prepared prior to reading the tobacco samples by using at least five standards from 5.0 to $100\,\mu g/ml$. This method is dependable for determining nornicotine solutions quantitatively, but it does not account for the presence of anabasine and anatabine in solution.

3.3.3 Thin-Layer Chromatography

Thin-layer chromatography is frequently used to separate pyridine alkaloids from tobacco extracts. This method is frequently used by tobacco breeders to screen plant populations and individual tobacco plants to insure seed purity. Thin-layer is considerably faster than paper chromatography and can also be used as a preliminary cleanup for more sensitive methods for analytical analysis such as capillary gas chromatography (GC). Semiquantitative analyses are obtained by developing color for individual alkaloids separated from tobacco extracts. Individual samples are separated on thin-layer strips or a number of samples can be spotted 1 cm apart in a line 1 cm from the margin of a thin-layer plate; a standard or a tobacco sample is spotted on either side of the plate for reference. The sample in question is spotted between two samples for comparison. Prefabricated plates that are uniformly coated and ready for use can be obtained in several dimensions from suppliers. Silica Gel G is the plate of choice for chromatography of tobacco extracts.

Plates are activated in a dry uncontaminated oven at 110 °C for 30 min and cooled to room temperature for 30 min before use. Samples are spotted on the plate with 10 µl capillary tubes and the plates are developed in chloroform:methanol:ammonia (60:10:1) v/v/v (Hodgson et al. 1965; Fejér-Kossey 1967). The solvent is allowed to migrate in an ascending direction for 10, 15, or 20 cm, depending upon the degree of separation required. Plates are run in a second dimension in a second solvent system for further separation and the solvent of choice depends upon the performance of the first system (Hodgson et al. 1965; Fejér-Kossey 1967). Plates are allowed to air dry for 30 min or until the solvent disappears. Individual plates are sprayed with 2% *p*-aminobenzoic acid in 95% ethanol and dried. Plates are placed in a thin-layer jar saturated with cyanogen bromide vapor from crystals until individual alkaloids develop color. The optical density of each individual spot can be determined quantitatively using a thin-layer scanner. Scanning requires that a standard curve be determined using known concentrations of the alkaloid to be determined. Only two standard curves, nicotine and nornicotine, are needed, since secondary alkaloids can be quantitatively determined from nornicotine. If screening samples for nornicotine, the chromatograms can be judged by the intensity of the spots, and plants that are converters can be identified. After the plates are developed, they can be covered with plastic and stored in a freezer for future reference.

3.3.4 Gas Chromatography

The use of gas chromatography (GC) for tobacco alkaloids analysis has been used by many investigators. This method gives the investigator the advantage of separation, response, and speed of analysis. Early GC methods used columns packed with numerous column packing materials heavily coated with many different liquid phases that gave long retention times and poor separation. Several liquid substrates were found that improved separation and reduced retention times. Weeks et al. (1969) separated nicotine, nornicotine, anabasine, anatabine, and myosmine. Five percent SE-30 (methyl silicone rubber), 10% Versamid 900 (an ethylenediaminelinoleic acid polyamide resin), and 10% DC 550 (methylphenyl silicone oil) were used to coat acid-washed and dimethyldichlorosilane-treated 60–80 mesh Chromosorb-W. Column lengths were 2.44 m coiled glass with 3.5-mm i.d. These columns gave separation of the five alkaloids with effective plate values of 500 to 1800 at a column temperature of 170 °C isothermal. Limit of detection and separation ranged from 0.5 μg/μl to 3 μg/μl. Fresh green tissue and cured tobacco samples were analyzed using the Versamid 900 column with limits of detection as low as 0.5 g of green tissue and 25 mg of cured tissue.

Capillary columns have become the state of the art for GC analysis of tobacco alkaloids (Lauterbach and Walker 1986). Instruments are equipped with flame ionization detectors (FID) and nitrogen-phosphorous detectors (NPD) using column lengths varying from 30 m to 60 m in length and internal diameters from 0.28 mm to 0.75 mm. The sensitivity range of detectors are from 0.05 to 1 μg/μl from which 1 μl of sample is injected into the GC. The column oven is operated isothermal or temperature programmed depending upon the desired analysis. Megabore columns (0.75 mm i.d.) are operated without a splitter or can be used as packed columns by adapting a makeup gas system to the detector. Megabore columns are advantageous to packed columns because they can handle larger samples with the resolving power of capillary columns giving good sensitivity. Apolar (SE-52, SE-30, DB-5, DC-550) or moderately polar capillary columns are the columns of choice for tobacco alkaloid analyses because they are more compatible with the solvents used in alkaloid extractions, compared to polar columns, and they are stable for long-term use. Polar columns (Carbowax 20 M) must be treated with a base and when an FID is used the peaks have a tendency to tail. This does not seem to be the case with an NPD when using KOH-treated Carbowax columns (Severson et al. 1981). A separation obtained from a flue-cured tobacco converter on a 0.75 mm i.d., 30 m SE 30 megabore column is shown in (Fig. 2). This sample contained 3.2% total alkaloids, of which 23% of the total alkaloids were secondary alkaloids. Myosmine was not detected.

Gas chromatographic analysis of tobacco alkaloids does not require derivation. The general procedures for GC analyses are as follows: (1) use the smallest sample applicable for analysis; (2) utilize preparations that clean up the sample without loss of alkaloids; (3) pre-extract the sample prior to alkaloid extraction with hexane (removes pigments and lipids); (4) extract alkaloids from tobacco with an aqueous acid solution, filter, partition the aqueous acid extract with organic solvent (hexane, methylene chloride, chloroform, etc.); (5) increase pH of the aqueous extract to pH 10 (5 N NaOH) and partition the basic solution with chloroform; (6) dry the chloroform solution over Na_2SO_4; and (7) concentrate the sample or analyze as is.

Standardization of GC parameters (injector temperatures, linear gas flows, column temperature, detector temperature) are required before analysis is performed. Detector response should be determined for each

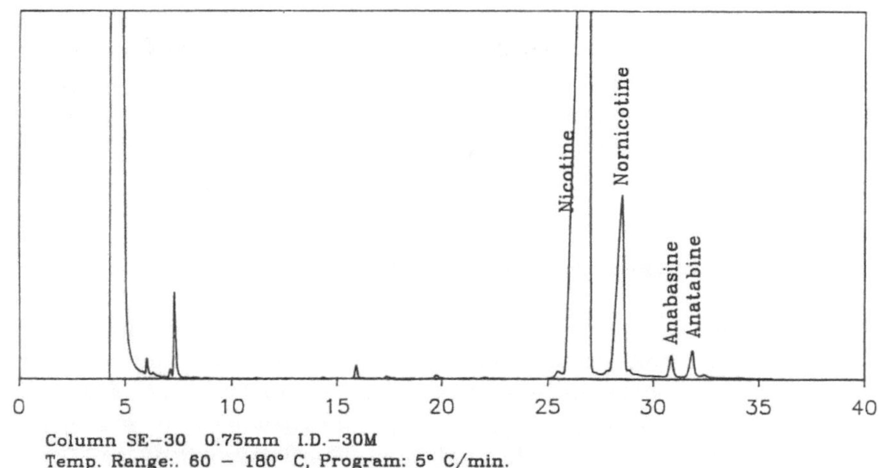

Fig. 2. Gas chromatogram of tobacco alkaloids from a flue-cured tobacco converter

compound in the sample to be analyzed. The internal standard method
should be employed if the GC is equipped with a data system and the
sample is automatically integrated. This method employs a standard that has
a similar molecular weight and polarity to the pyridine alkaloids. It should
preferably elute prior to the compounds to be analyzed and not interfere
with resolution or interact with the sample in any manner. It is also necessary
to establish the linear concentration of the internal standard, as well as
for each alkaloid compound in the sample, and to determine the relative
response factor of the internal standard with each compound. Anethole and
quinoline are two compounds commonly used as internal standards in GC of
tobacco alkaloids.

Flue-cured tobacco total alkaloids are frequently analyzed by GC using
an autosampler and an automatic injector. A small (10–25 mg) tobacco
sample must contain sufficient alkaloid concentrations in order to justify an
autosampler. A rapid and easy extraction procedure is used to prepare a
large number of samples for analysis. Severson et al. (1981) developed an
extraction technique for an automated GC method using glass capillary
columns (60 m × 0.28 i.d. Carbowax 20 M, ethylenediamine-treated) and a
nitrogen-specific detector. Finely ground tobacco samples (10–25 mg) were
weighed into 4-ml Hewlett Packard sample vials. One ml of extracting
solution (2% KOH in MEOH) was added to each sample and sample vials
were sealed with a crimping tool. Samples were allowed to extract over
night or were placed in a sonic bath for 15 min. The samples are run using a
60:1 split-injection ratio. Temperature was 200 °C for the injector port and
280 °C for the nitrogen-phosphorus (NPD) flame ionization detector. The
NPD detector is operated according to the manufacturers recommendations.

The oven temperature is programmed from 170 to 200 °C at 4 °C/min. Column linear gas flows were calibrated to 20 cm/s and analysis was complete after 15 min. Anabasine and myosmine were not completely resolved. Myosmine was not present in many samples and usually occurs in concentrations too low to be detected. This method is applicable to green or cured samples.

3.3.5 High Performance Liquid Chromatography

The use of HPLC for separation of tobacco alkaloids has been limited to few studies (Court 1986). Saunders and Blume (1981) developed a procedure for analyses of nicotine, nornicotine, anabasine, and anatabine in tobacco. This method utilized a μBondapak C_{18} reversed-phase column. The mobile phase of 40% methanol containing 0.2% phosphoric acid was adjusted to pH 7.25 with triethylamine. Retention times with 0.5 ml/min flow rates were 23 min for nicotine, 13.9 min for anatabine, 12.2 min for anabasine, and 11.0 min for nornicotine. This method is capable of analyzing samples containing 40–50 ng of alkaloid, and quantitative results were reliable. The column was stable for 800–1000 injections before it showed appreciable degradation in separation but was still capable of separating nicotine and nornicotine.

Both green and dried samples were used in the above method. Dried samples are ground to pass a 2-mm screen. A 0.5-g sample was extracted for 24 h with constant agitation with 10 ml of 25 mM sodium phosphate buffer (pH 7.8) and filtered under reduced pressure through Whatman #2 filter paper. Each extract was filtered through a 0.45-μm Millipore filter and sealed in a screw cap vial for automatic injection. One-gram fresh green samples were ground in 10 ml of 40% methanol (v/v) containing 0.1% (v/v) 1 N HCL with a homogenizer, centrifuged at 500 g for 3 min, and filtered through Whatman #2 filter paper under vacuum. Each sample was diluted one to four and filtered through a 0.45-μm Millipore filter prior to automatic injection. There is no reason to pre-prep fresh tissue to remove chlorophyll or other interfering substances. With the use of an automatic injection system, the procedure is applicable to screening individual plants from a breeding program. In the case of green plants, one leaf could be assayed for alkaloids and seed could still be collected from the plant if the analysis indicated that the plant was a nonconverter.

3.4 Cundiff-Markunas Extraction Method

Alkaloids can be determined from tobacco after extraction by titration. This method determines total alkaloids (as nicotine), tertiary alkaloids (as nicotine), and total secondary alkaloids (as nornicotine). Titration is sensitive and accurate. Compared to the more sensitive methods used routinely for alkaloid analyses, a larger sample and longer extraction time are required, but accurate analytical data may be efficiently obtained (Cundiff and Markunas 1955).

A ground sample is weighed accurately to 0.8 g and placed in an 125-ml Erlenmeyer flask. Ten ml of 1 N HCL are added as a wetting agent, 10 ml of 5 N NaOH are added to increase alkalinity, and 50 ml of chloroform are added to the flask. The flask is stoppered and placed on a shaker for 20 min. Two to 3 g of analytical grade celite are added to absorb water, the sample is agitated briefly, and filtered through S&S #605 filter paper (a rapid quantitative paper). Five ml aliquots are added to two calibrated test tubes. One tube is gently aerated to remove ammonia, and 0.5 ml of acetic anhydride (reagent grade) is added. The sample is allowed to stand for 10 min before titrating. One drop of 0.2% crystal violet indicator (prepared by adding 1.6 g crystal violet to 800 ml of $CHCl_3$) is added to each tube with an eye dropper. The samples are titrated with 0.01 N perchloric acid (prepared by mixing 70% $HClO_4$ with glacial acetic acid). The perchloric acid ($HClO_4$) solution is added from a buret while mixing constantly with a Vortex mixer until the end point (a dark green color) is reached. The second tube, treated with acetic anhydride, is titrated in the same manner. Percent total alkaloids (as nicotine), percent tertiary alkaloids (as nicotine), and percent total secondary alkaloids (as nornicotine) are calculated using the following equations:

% total alkaloids = $(V_1 \times N \times 81.12)/g$ sample weight

% tertiary alkaloids = $[(2V_2 - V_1) \times N \times 81.12]/g$ sample weight

% total secondary alkaloids = $[2(V_1 - V_2) \times N \times 74.1]/g$ sample weight,

where V_1 = ml of 0.01 N $HClO_4$ for nonacetylated aliquot, V_2 = ml of 0.01 N $HClO_4$ for acetylated aliquot, N = normality of $HClO_4$, 81.12 = ½ the molecular weight of nicotine, and 74.1 = ½ the molecular weight of nornicotine. Other secondary alkaloids, anabasine and anatabine, also contributing to the concentration of secondary alkaloids (Cundiff and Markunas 1960; Cundiff and Markunas 1964), are measured as nornicotine when using this method.

3.5 Alternative Methods

A large number of varied analytical procedures have been developed for the determination of alkaloids in tobacco. Only a few of these procedures have been described in detail. Other methods available for the analysis of alkaloids include near-infrared reflectance spectroscopy (NIRS) (McClure and Williamson 1986), lipophilic gel chromatography (Snook and Chortyk 1986), radioimmunoassay (Castro et al. 1979), mass spectrometry (Green et al. 1980), and pyrolysis-gas chromatography (Rosa 1979).

4 Summary

There are a wide variety of methods available for the analysis of alkaloids in tobacco. Each method is associated with certain advantages and disadvantages, but fills a certain niche for tobacco scientists. If the percent total alkaloids determined as nicotine suffices, then steam distillation or auto-analyzer are appropriate methods. When secondary alkaloids need to be quantitatively determined, then one of the chromatographic methods is more appropriate.

Acknowledgments. The authors gratefully acknowledge Drs. H.R. Burton and J.F. Chaplin for their critical review of the manuscript. We also thank Ms. J. Jwong for her preparation of the figures.

References

Akehurst BC (1981) Tobacco, 2nd edn. Longman, New York
Bush LP (1981) Physiology and biosynthesis of tobacco alkaloids. Rec Adv Tob Sci 7:75–106
Campbell CR, Chaplin JF, Johnson WH (1982a) Cultural factors affecting yield, alkaloids, and sugars of close-grown tobacco. Agron J 74:279–283
Campbell JS, Chaplin JF, Boyette DM, Campbell CR, Crawford CB (1982b) Effect of plant spacings, topping heights, nitrogen rates, and varieties of tobacco on nicotine yield and concentration. Tob Sci 26:66–69
Castro A, Monji N, Malkus H, Eisenhart W, McKennis H Jr, Bowman ER (1979) Automated radioimmunoassay of nicotine. Clin Chim Acta 95:473–481
Chaplin JF, Miner GS (1980) Production factors affecting chemical components of the tobacco leaf. Rec Adv Tob Sci 6:3–63
Court WA (1986) High performance liquid chromatography of tobacco and tobacco smoke components. Rec Adv Tob Sci 12:143–184
Cundiff RH, Markunas PC (1955) Determination of nicotine, nornicotine, and total alkaloids in tobacco. Anal Chem 27:1650–1653
Cundiff RH, Markunas PC (1960) Modification of the extraction procedure for determination of alkaloids in tobacco. J Assoc Off Agric Chem 43:519–524
Cundiff RH, Markunas PC (1964) Abbreviated techniques for determinations of alkaloids in tobacco using the extraction procedure. Tob Sci 8:136–137
Dawson RF (1945) An experimental analysis of alkaloid production in *Nicotiana*: the origin of nornicotine. Am J Bot 32:416–423
Enzell CR, Wahlberg I, Aasen AJ (1977) Isoprenoids and alkaloids of tobacco. Prog Chem Org Nat Prod 34:1–79
Fejér-Kossey O (1967) The separation of ten tobacco alkaloids by thin-layer chromatography. J Chromatogr 31:592–593
Glock E, Wright MP (1963) Determination of nornicotine in tobacco and smoke by the 1,3-indanedione spectrophotometric method. Anal Chem 35:246–251
Green CR, Colby DA, Cooper PJ, Heckman RA, Lyerly LA, Thome FA (1980) Advances in analytical methodology of leaf and smoke. Rec Adv Tob Sci 6:123–183
Griffith RB (1957) The rapid determination of total alkaloids by steam distillation. Tob Sci 1:130–137
Griffith RB, Jeffrey RN (1948) Improved steam distillation apparatus. Application to determination of nicotine in green and dry tobacco. Anal Chem 20:307–311

Harvey WR, Stahr HM, Smith WC (1969) Automated determination of reducing sugars and nicotine alkaloids on the same extract of tobacco leaf. Tob Sci 13:13–15

Hawks SN Jr, Collins WK (1983) Principles of flue-cured tobacco production. SN Hawks Jr, WK Collins, Raleigh, NC

Hodgson E, Smith E, Guthrie FE (1965) Two dimensional thin-layer chromatography of tobacco alkaloids and related compounds. J Chromatogr 20:176–177

Jeffrey RN, Eoff WH (1955) Paper chromatographic method for determining alkaloids in tobacco. Anal Chem 27:1903–1904

Jeffrey RN, Tso TC (1955) Qualitative differences in the alkaloid fraction of cured tobaccos. J Agric Food Chem 3:680–682

Lauterbach JH, Walker KM (1986) Recent advances in gas chromatography and their applications in tobacco research. Rec Adv Tob Sci 12:185–236

Long RC, Woltz WG (1977) Environmental factors affecting chemical composition of tobacco. In: McKenzie JL (ed) Recent advances in the chemical composition of tobacco and tobacco smoke. Proc Am Chem Soc Symp, Am Chem Soc, Agricultural and Food Chemistry Division, pp 116–163

Mann TJ, Matzinger DF, Wernsman EA (1975) Genetic control of tobacco constituents. Tob Res 1:1–12

McClure WF, Williamson RE (1986) Status of near-infrared technology in the tobacco industry. Rec Adv Tob Sci 12:3–53

Mosley JM, Rayburn CH (1957) Nornicotine-type alkaloids as a factor in the taste and composition of cigarette smoke. Abstr. 11th Tobacco Chemists' Research Conference, New Haven, CT

Mulchi CL (1985) Environmental factors affecting the growth, chemistry, and quality of tobacco. Rec Adv Tob Sci 11:3–46

Rosa N (1979) Pyrolysis-gas chromatographic estimation of tobacco alkaloids and neophytadiene. J Chromatogr 171:419–423

Sadler WW, Chesson RR, Schoenbaum AW (1960) Automated procedure for determining the nicotine content of steam distillates. Tob Sci 4:208–212

Saunders JA, Blume DE (1981) Quantification of major tobacco alkaloids by high-performance liquid chromatography. J Chromatogr 205:147–154

Severson RF, McDuffie KL, Arrendale RF, Gwynn GR, Chaplin JF, Johnson AW (1981) Rapid method for the analysis of tobacco nicotine alkaloids. J Chromatogr 211:111–121

Sisson VA, Severson RF (1990) Alkaloid composition of the *Nicotiana* species. Beitr Tabakforsch 14:327–339

Snook ME, Chortyk OT (1986) Advances in lipophilic gel chromatography of tobacco and tobacco smoke components. Rec Adv Tob Sci 12:237–297

Stephens RL, Weybrew JA (1959) Isatin: a color reagent for nornicotine. Tob Sci 3:48–51

Tso TC (1990) Production, physiology, and biochemistry of tobacco plant. IDEALS, Beltsville, MD

Tso TC, Kasperbauer MJ, Sorokin TP (1970) Effect of photoperiod and end-of-day light quality on alkaloids and phenolic compounds of tobacco. Plant Physiol 45:330–333

United States Department of Agriculture, Foreign Agricultural Service (1991) World tobacco situation. Circ Ser FT-12-91, USDA-FAS, Washington, DC, pp 48–49

van Bevel CHM (1953) Chemical composition of tobacco leaves as affected by soil moisture conditions. Agron J 45:611–614

Weeks WW, Davis DL, Bush LP (1969) Gas chromatographic separation of tobacco alkaloids. J Chromatogr 43:506–509

Weybrew JA, Wan Ismail WA, Long RC (1983) The cultural management of flue-cured tobacco quality. Tob Sci 27:56–61

Williams S (1984) Official methods of plant analysis, 14th edn. Association of Official Analytical Chemists, Arlington, VA, pp 62–63

Assessment of Burley and Dark Tobacco Alkaloids During Storage, Aging, and Fermentation

R.A. Andersen

1 Importance of Pyridine Alkaloids to Tobacco Quality and Usability

Pyridine alkaloids in tobacco (*Nicotiana tabacum* L.) and tobacco smoke are widely recognized for their contributions to tobacco quality and usability (Palmer 1963; Tso 1990). Nicotine is the most abundant alkaloid and may provide a pleasurable alerting effect associated with tobacco uses. The neurobiology of nicotine has been reviewed by Ashton et al. (1979). Health risks of tobacco alkaloids have also been summarized (Davis 1987). Nitrosated alkaloids may be carcinogenic (Hoffmann et al. 1987). Sums of alkaloids other than nicotine are generally small but may range up to 20% of total alkaloids (Andersen et al. 1991). Most alkaloids identified in postharvest burley and dark tobaccos or products containing them are given in Table 1. All of these compounds may be referred to as alkaloids because they occur naturally in green or processed tobacco, or they can be classified as "parent" alkaloids and their acylated, nitrosated, or oxidized derivatives. Structures of six representative tobacco alkaloids are shown in Fig. 1.

2 Importance of Storage, Aging, and Fermentation

Storage followed by aging or fermentation generally begins where curing ends. Burley and dark tobaccos are stored for about 1 month before redrying and a required aging or fermentation. Fermentation is accompanied by a vigorous chemical change, whereas aging is a slow, mild fermentation. Prior to aging, air-cured burley and dark tobaccos are taken to 10 to 12% moisture by a redrying process. Air-cured burley leaf is generally aged for about 2 years. Fermentation, often called "sweating", is usually used for cigar tobaccos.

Outstanding effects of aging or fermentation are development of the desired aroma, elimination of harshness, improvement of the bitter taste, and disappearance of green shades. The leaf acquires a darkened color, diminished luster and stickiness, improved combustibility, and reductions of irritating qualities in its smoke.

Modern Methods of Plant Analysis, Volume 15
Alkaloids (ed. by Linskens/Jackson)
© Springer-Verlag Berlin Heidelberg 1994

Table 1. Alkaloids identified in postharvest burley and dark tobaccos or products containing them

"Parent"	Acylated (acronym)	Nitrosated (acronym)	Oxidized ("alternate name")
Nicotine	Formylnornicotine (FNN)	Nitrosonornicotine (NNN)	Cotinine
Nornicotine	Formylanatabine (FAT)	4-(Methylnitrosamino)-1-(3-pyridyl-)-1-butanone (NNK)	2,3'-Dipyridyl ("bipyridyl")
Anatabine	Acetylnornicotine (ANN)		
Anabasine	Acetylanatabine (AAT)	Nitrosoanatabine (NAT)	
Myosmine	n-Butanoylnornicotine (BNN)	Nitrosoanabasine (NAB)	
Ethylnornicotine	n-Hexanoylnornicotine (HNN)	4-(Methylnitrosamino)-4-(3-pyridyl)-butanal (NNA)	
Isopropylnornicotine	n-Octanoylnornicotine (ONN)	4-(Methylnitrosamino)-1-(3-pyridyl)-1-butanol (NNAL)	
		4-(Methylnitrosamino)-1-(3-pyridyl)-1-butanol (iso-NNAL)	
		4-(Methylnitrosamino)-4-(3-pyridyl) butyric acid (iso-NNAC)	

3 Chemical Analyses[1]

3.1 Total Tobacco Alkaloids

Total alkaloids in tobacco are commonly determined spectrophotometrically. The Griffith Still Method (Griffith 1957) utilizes a steam distillation apparatus based on that of Willits et al. (1950); absorbance is measured at 236, 259, and 282 nm. Automated methods such as those of Harvey et al. (1969) and Davis (1976) are more commonly used today. Pyridyl moieties react with cyanogen bromide and aniline to form a chromophore measurable at 460 nm.

The method for total alkaloids (as nicotine) used in the Agronomy Department, University of Kentucky, is based on that of Harvey et al.

---

[1] Trade names are given as part of the analytical conditions and not as an endorsement of products.

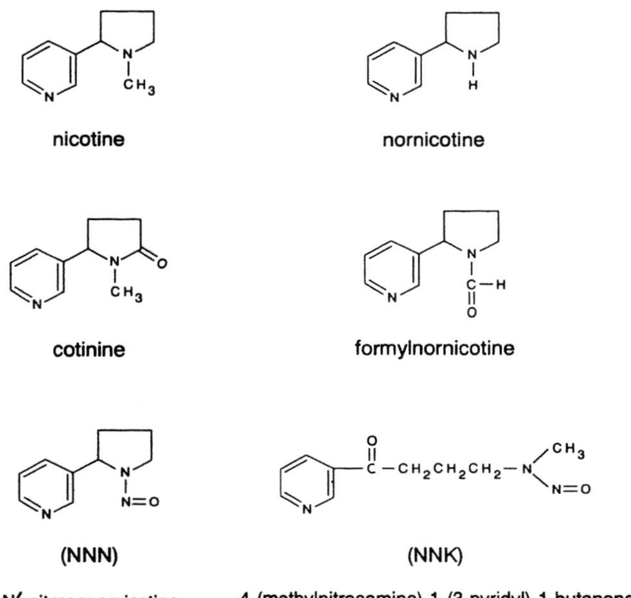

Fig. 1. Six common tobacco alkaloids or alkaloid derivatives

(1969) and employs an AutoAnalyzer with manifold described in Technicon Industrial Method No. 117-71A, Bran + Lubbe, Buffalo Grove, IL (Fig. 2).

Details are:

Weigh 100 to 200 mg of ground tobacco with known moisture into a 25 × 200-mm tube. Add 250 mg charcoal and 100 ml extracting solution. Shake tube contents 10 min. Filter through Whatman #40 filter paper. Sample the filtrate in AutoAnalyzer (fume hood). Determine nicotine alkaloids by interpolation from a standard curve.

Reagents are:

Extracting solution. Add 200 ml methanol and 220 ml 0.1 N HCl in a 1-l volumetric flask. Mix and dilute to mark with H_2O.

Buffered aniline. Dissolve 11.4 g anhydrous disodium hydrogen phosphate and 4.4 g citric acid in 800 ml H_2O. Add 3 ml aniline and 200 ml 95% ethanol. Mix and add 0.5 ml Brij-35 (Fisher, Pittsburgh, PA).

Cyanogen bromide. Dissolve 20 g cyanogen bromide in 800 ml H_2O. Add 200 ml 95% ethanol. Mix and add 0.5 ml Brij-35 and store in a glass bottle. Caution – cyanogen bromide vapors are toxic; prepare reagent in well-ventilated hood!

The standards are:

Stock standard, 1 mg/ml. Dissolve 0.307 g nicotine hydrogen tartrate dihydrate (or equivalent amount of redistilled nicotine) in 100 ml extracting solution.

Working standards. Pipet 1 to 5 ml stock standard into 100-ml volumetric flask and dilute to volume with extracting solution (a 1-ml aliquot of diluted stock standard is equivalent

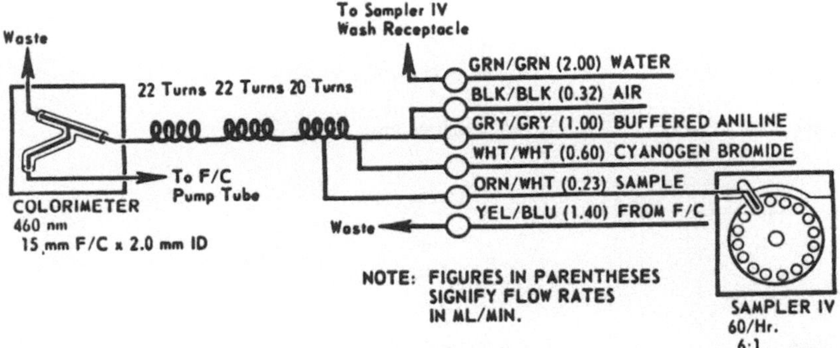

Fig. 2. Manifold for total alkaloids in tobacco extracts, 0–5 mg % total alkaloids (as nicotine)

to 1 mg % nicotine). Multiple working standards are used to establish linearity. A 3 mg % standard is recommended for day-to-day instrument calibration.

3.2 Individual Tobacco Alkaloids

Pyridine alkaloids in tobacco are usually determined by gas chromatography. The "parent" alkaloids – nicotine, nornicotine, anatabine, anabasine, and myosmine – may be determined in a single extract. A capillary GC procedure we use (Andersen et al. 1990) is based on that of Severson et al. (1981) but with modification.

Details are:

A 0.5-g tobacco sample is vigorously shaken for 1 min with a mixture of 5 ml of 1.0 N NaOH, 30 ml ethyl acetate, and 0.5 ml of an azobenzene solution (2 mg/ml in ethyl acetate), then set aside for 1 h. A 3-µl aliquot of the top layer is injected into a Hewlett-Packard Model 5880A GC in splitless mode utilizing a 30 m × 0.32-mm fused silica HB-5 nonpolar capillary column (1-µ film) with temperature programmed from 100 °C at 4 °C per min (following a 1-min hold at 100 °C) to 250 °C. Quantifications are by internal standardization with azobenzene after calibration of retention times and response factors for FID with authentic alkaloids. Response factors are determined by these steps: (1) detector responses are plotted versus six or more amounts of a standard compound ranging from 0–2000 ng that are injected into the GC, (2) curvilinear regression is used to determine the best mathematical function to describe the resultant empirical curve; in our experiments this was the quadratic model $y = a + bx + cx^2$, where y = detector response units, x = amount of standard compound, and a, b, and c are coefficient constants, and (3) a response factor is obtained by taking the derivative of the mathematical model.

Quantitative results obtained during storage, aging, or fermentation are expressed as µg/g dry wt and because of dry weight changes may be normalized on the basis of calcium concentration at the start of storage.

Nitrosated, acylated, and oxidized alkaloids are extracted and separated from the bulk of the "parent" alkaloids by liquid-liquid partitioning at pH 5 and may be determined by a GC procedure described by Andersen et al. (1990). "Parent" alkaloids are generally more basic than their derivatives and do not readily partition into the organic phase at pH 5. Acylated, nitrosated, and oxidized derivatives (Table 1) may be determined in a single GC run.

The procedure is:

A 1-g sample of tobacco is extracted for 1 h with constant shaking at room temperature with 10 ml of 0.025 M citrate–0.05 M phosphate buffer, pH 4.5, containing 5 mM ascorbic acid. After extraction, the mixture is adjusted to pH 5 and partitioned once with 30 ml ethyl acetate. A 20-ml aliquot of the organic phase is partitioned three times with 5-ml portions of 1 N HCl. Combined aqueous phases are adjusted to pH 5 with 10 N NaOH in an ice bath and then partitioned three times with 5-ml portions of chloroform. Sodium sulfate is added to the combined chloroform phases, which are stored in sealed containers at 5 °C until analysis by GC or GC-MS. Aliquots (0.5–4 µl) of the final chloroform solution are injected into a Hewlett-Packard Model 5880A GC under these conditions: operation in splitless mode; 30 m × 0.25 mm fused silica DB-5 nonpolar capillary column (1-µ film) and thermionic N-P detector; inlet 280 °C; inlet purge flow 1–2 ml/min; purge time 1 min; helium carrier linear velocity 31 cm/s; column temperature programmed from 60 °C with initial 1-min hold to 260 °C at 4 °C/min and then held at 260 °C for 5 min; detector temperature 280 °C.

Quantifications are by internal standardization with azobenzene after calibration of retentions (see Table 2) and nitrogen-phosphorus detector (NPD) response factors for authentic compounds. Response factors are determined by these steps:

(1) NPD responses plotted versus seven or more concentrations of a standard compound ranging from 0–1500 ng injected into GC, (2) curvilinear regression used to determine best fit mathematical function describing empirical curve; in our experiments this was the power model $y = ax^b$, where y = detector response units, x = amount of standard compound, a = coefficient constant, and b = exponent constant (the a and b power model constants were calculated for a given detector bead), and (3) response factor obtained by taking the derivative of the mathematical value for a given bead response.

Table 2. GC indices of some nitrosated and acylated N'-substituted pyridine alkaloids identified in burley tobacco

Compound	Kovats index[a]
N'-Nitrosonornicotine (NNN)	1756
N'-Formylnornicotine (FNN)	1783
N'-Nitrosoanatabine (NAT)	1808
N'-Formylanatabine (FAT)	1858
N'-Acetylnornicotine (ANN)	1827
N'-Acetylanatabine (AAT)	1902
4-(N-Methyl-N-nitrosamino)-1-(3-pyridyl)butanone (NNK)	1935
N'-n-Butanoylnornicotine (BNN)	1961
N'-n-Hexanoylnornicotine (HNN)	2159
N'-n-Octanoylnornicotine (ONN)	2367

[a] Index for the DB-5 capillary column used with conditions described in Section 3.2.

Corrections are made for recoveries of individual compounds carried through the entire procedure; recoveries ranged from 16 to 108%.

Another capillary method specific for nitrosated pyridine alkaloids utilizes a thermal energy analyzer (TEA) detector (Peng 1990). The procedure for a 1-g sample is similar to that described above using an NPD except for the following:

The final chloroform extracts are filtered through glass wool, taken to near dryness on a rotating evaporator followed by removal of all solvent with a stream of nitrogen. The residue is dissolved in 50–100 µl acetone solution containing 3.39 µg/ml N-nitrosoguvacoline as internal standard. A 3- to 5-µl aliquot is injected into a Hewlett Packard (HP) 5890 GC in splitless mode with He carrier at 1.1 ml/min and injector temperature 230 °C. A 50-m nonpolar GB-5 fused silica capillary column is programmed from 40 °C (held 4 min) to 200 °C at 30 °C/min, then held at 200 °C for 27 min. An HP 3396A integrator is used to quantify the signal from a TEA Model 543 analyzer (Thermedics Inc.). The TEA pyrolyzer and interface is maintained at 500 and 250 °C, respectively. Vacuum with and without ozone is about 0.3 and 0.7 mmHg, respectively, and attenuation is 4.

Quantification of nitroso-alkaloids is by internal standardization. Calibration of retention times and response factors is made with authentic nitrosamines.

A packed column GC method for nitrosated pyridine alkaloids in tobacco (Spiegelhalder et al. 1989) allows simultaneous determinations of nitrate and nitrite by continuous flow spectrophotometry. The procedure includes aqueous alkaline extraction of tobacco in the presence of an antibacterial agent. Nitrosamines are cleaned up by liquid-liquid extraction and removal of nicotine. Temperature-programmed GC separations on 10%-OV 17 on chromosorb WHP (80/100 mesh) are carried out. The analysis utilizes N′-nitrosopentylpicolylamine or N-nitrosodibenzylamine as internal standards and TEA detection.

3.3 Nitrite

Nitrite content is important in the assessment of tobacco and alkaloids during storage and postharvest treatments because it is related to nitrosamine accumulation and bacterial activity. A colorimetric method for nitrite in tobacco samples has been developed that stabilizes and decolorizes leaf extracts for automated continuous flow analysis at ppb or greater (Crutchfield and Burton 1989). It avoids the problem that nitrite measurement in unsterilized tissue is error-prone because of bacteria-mediated reduction of nitrate to nitrite. The manifold arrangement is in Fig. 3.

Procedural details are:

Weigh 0.500 g plant material into 25 × 200 mm test tubes. Add 50 ml of extraction solution, stopper, and shake for 30 min at high speed on reciprocal shaker. Filter through Whatman #1 filter paper into 16 × 125 mm plastic test tubes, discarding first 5–10 ml of filtrate. Collect 10 ml and add 0.5 g of Norit A. Stopper and shake 15 min to decolorize. Allow the carbon to settle, then filter through Whatman #42 filter paper into sample cups. Analyze on the Technicon (Bran + Lubbe) Auto-Analyzer.

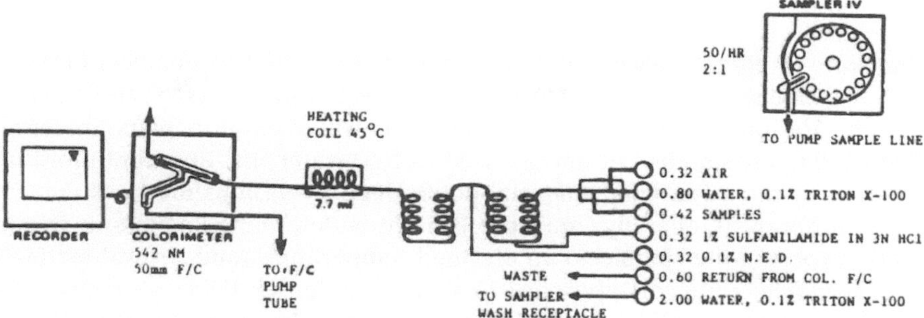

Fig. 3. Nitrite analysis manifold arrangement using a Technicon Auto-Analyzer System II

Standards should be prepared each day from 50 ppm nitrite stock using fresh extraction solution. To determine a baseline reference point, a portion of extraction solution is included with each set of standards.

Reagents and standards are:

Extraction solution. 1% KCl, 0.5% sulfanilamide, 0.1% Triton X-100 (10 drops/l) in deionized water.

Standard stock solution. 50 ppm NO_2-N in deionized water.

Working standards. 20–100 ppb NO_2-N. Place 0.5 ml aliquot of 50 ppm stock solution in 250-ml vol. flask and dilute to volume with extraction solution to make the 100 ppb standard. Place 10, 20, 30, 40 ml of this standard in 50-ml volumetric flasks and dilute each to volume with extracting solution to make 20, 40, 60, 80 ppb standards. (Standards and extraction solution must be made fresh daily.)

Color reagents. 1% sulfanilamide in 3N HCl. 0.1% N-(1-naphthyl)-ethylenediamine dihydrochloride in deionized water. (Use no wetting agents.) Sampler wash solution. Deionized water with 10 drops/l of Triton X-100.

3.4 Calcium

In postharvest studies of tobacco, initial dry weight is often the basis for calculation and evaluation. However, there is usually dry matter loss from biochemical change during the observation period. It is therefore advisable to determine corrected weights based on the concentrations of inert components such as calcium (Frankenburg 1950). The method for calcium by atomic absorption as carried out in the Agronomy Department of the University of Kentucky follows:

Digest a 0.25-g tobacco sample at 80 °C overnight in 8.0 ml of nitric acid: perchloric acid = 9:1. Continue heating the mixture to fumes and take to dryness. Dissolve the residue in 25 ml of 1 N HCl. Dilute to appropriate concentration with distilled water. Analyze the solution at 422.7 nm in an atomic absorption spectrophotometer (Instrumentation Laboratory S11) using a nitrous oxide-acetylene flame.

3.5 Moisture

Moisture in cured tobacco may predispose the material to significant chemical and biochemical changes during postharvest storage, aging, and fermentation. Various physical states or forms of water exist in plant material. Depending on amounts of energy used to free water, the apparent moisture content of the same plant material determined by various methods may be quite different. Generally, moisture content is determined as loss in weight upon exposing the sample to an elevated temperature, and several methods are available using this approach (Tso and Andersen 1974). A convenient method for moisture in tobacco utilizes the Rapid Moisture/Volatiles Tester available from C.W. Brabender Inst., Inc. (S. Hackensack NJ 07606). The instrument features a built-in analytical balance. The recommended procedure for tobacco requires drying for only 20 min at 130 °C. Another method is freeze drying. Errors in moisture content may occur because of sample decomposition or loss of volatile compounds during heating or under conditions of reduced pressure.

References

Andersen RA, Fleming PD, Burton HR, Hamilton-Kemp TR, Hildebrand DF, Sutton TG (1990) Levels of alkaloids and their derivatives in air- and fire-cured KY 171 dark tobacco during prolonged storage: effects of temperature and moisture. Tob Sci 34:50–56

Andersen RA, Fleming PD, Burton HR, Hamilton-Kemp TR, Sutton TG (1991) Nitrosated, acylated and oxidized pyridine alkaloids during storage of smokeless tobaccos: effects of moisture, temperature and their interactions. J Agric Food Chem 39:1280–1287

Ashton H, Marsh VR, Millman JE, Rawlins MD, Stepney R, Telford R, Thomson JW (1979) Patterns of behavioral, autonomic and electrophysiological response to cigarette smoking and nicotine in man. In: Remond H, Izard C (eds) Electrophysiological effects of nicotine. Elsevier/North Holland Biomedical Press, New York, pp 159–182

Crutchfield J, Burton HR (1989) Improved method for the quantitation of nitrite in plant materials. Anal Lett 22:555–571

Davis DL (1987) Tobacco use and associated health risks. Adv Behav Biol 31:15–23

Davis RE (1976) A combined automated procedure for the determination of reducing sugars and nicotine alkaloids in tobacco products using a new reducing sugar method. Tob Sci 20:146–151

Frankenburg WG (1950) Chemical changes in the harvested tobacco leaf. Part II. Chemical and enzymatic conversions during fermentation and aging. Adv Enzymol 10:325–441

Griffith RB (1957) The rapid determination of total alkaloids by steam distillation. Tob Sci 1:130–137

Harvey WR, Stahr HM, Smith WC (1969) Automated determination of reducing sugars and nicotine alkaloids on the same extract of tobacco leaf. Tob Sci 15:13–15

Hoffmann D, Adams JD, Lisk D, Fisenne I, Brunnemann KD (1987) Toxic and carcinogenic agents in moist and dry snuff. J Natl Cancer Inst 79:1281–1286

Palmer JK (1963) Changes in the nitrogenous constituents of burley tobacco during curing and aging. Tob Sci 7:93–96

Peng Q (1990) Alkaloids, nitrates, nitrites and tobacco specific nitrosamines in dark tobacco. Masters thesis, University of Kentucky, Lexington, KY

Severson RF, McDuffie KL, Arrendale RF, Gwynn GR, Chaplin JF, Johnson AW (1981) Rapid method for the analysis of tobacco nicotine alkaloids. J Chromatogr 211:111–121

Spiegelhalder B, Kubacki SJ, Fischer S (1989) A method for the determination of tobacco-specific nitrosamines (TSNA), nitrate and nitrite in tobacco leaves and processed tobacco. Beitr Tabakforsch Int 14:135–144

Tso TC (1990) Production, physiology, and biochemistry of tobacco plant. Ideals, Beltsville, MD

Tso TC, Andersen RA (1974) Analysis of leaf tobacco. In: Encyclopedia of industrial chemical analysis, vol 19. John Wiley, New York, pp 133–160

Willits CO, Swain ML, Connelly JA, Brice BA (1950) Spectrophotometric determination of nicotine. Anal Chem 22:430–433

Detection of Alkaloids
in Environmental Tobacco Smoke

M.W. Ogden and P.R. Nelson

1 Introduction

1.1 Tobacco and Tobacco Smoke

Tobacco is one of the most complex plants known, drawing its complexity from both genetic and environmental sources. Of the more than 60 genetic species of plants belonging to the genus *Nicotiana*, only two (*N. tabacum* and *N. rustica*) are widely cultivated for use as tobacco. *N. tabacum*, the tobacco of commerce, is a hybrid plant capable of a high degree of variability, as evidenced by the different major commercial types (flue-cured, burley, Maryland, and Oriental) and the numerous agronomic varieties within each type (Wolf 1967). Each of these types, in addition to others, is currently used in the manufacture of cigarettes, cigars, and pipe tobaccos.

Superimposed on the composition of the tobacco leaf constituents is a mixture of enormous complexity resulting from combustion of these tobaccos on smoking. One recent review estimates that at least 3875 individual components have been identified to date in tobacco smoke (Dube and Green 1982). Of these, 2549 are unique to smoke, with the remaining 1135 compounds being common to both tobacco leaf and smoke. Of the 3875 compounds known, 921 are N-heterocycles (Dube and Green 1982).

The smoke from burning tobacco is typically characterized as either mainstream or sidestream smoke. Mainstream smoke is that smoke emerging through the mouth end of the tobacco product during puffing and is primarily inhaled by the smoker. Sidestream smoke is principally that smoke emanating from the burning end of the tobacco product.

Environmental tobacco smoke (ETS) is more difficult to define precisely than is mainstream or sidestream smoke. A dynamic mixture of substances present in ambient air as a specific result of tobacco smoking, ETS consists primarily of aged and diluted sidestream smoke (ca. 85%) and exhaled mainstream smoke (ca. 15%). Exhaled mainstream smoke, as a component of ETS, is quantitatively different from inhaled mainstream smoke.

Three general processes occur simultaneously during tobacco combustion: pyrolysis, pyrosynthesis, and distillation. Pyrolysis refers to thermal decomposition of the organic leaf components into smaller molecules. Being partially unstable chemically, some of these newly formed fragments can recombine to form additional components (pyrosynthesis). Both pyrolysis and pyrosynthesis result in the presence of chemical species in tobacco

Modern Methods of Plant Analysis, Volume 15
Alkaloids (ed. by Linskens/Jackson)
© Springer-Verlag Berlin Heidelberg 1994

smoke that were not originally present in the tobacco. The third process, distillation, results in the intact transfer of certain compounds directly from tobacco to smoke. The principal alkaloid of tobacco, nicotine, contributes to the complexity of tobacco smoke through all three processes.

1.2 Tobacco-Related Alkaloids

Prior to embarking on a discussion of tobacco-related alkaloids, it appears warranted to outline briefly what we consider an *alkaloid* to be, since there is considerable disparity of opinion (Pelletier 1983). We concur with the broader definition offered by Pelletier: "An alkaloid is a cyclic compound containing nitrogen in a negative oxidation state which is of limited distribution among living organisms" (Pelletier 1983). Accordingly, we consider all of the N-heterocycles of tobacco and smoke to be alkaloids, regardless of molecular structure complexity or confirmed significant pharmacological activity.

A listing of the tobacco alkaloids is well beyond the scope of this chapter (see Chaps. 7 and 8) and would serve no useful purpose in the context of the present topic. Several reviews and historical surveys are cited for the interested reader (Kuhn 1965; Pailer 1965; Wynder and Hoffmann 1967; Stedman 1968; Leete 1983). In this chapter, all discussion will be restricted to the most abundant pyridine alkaloids in tobacco and selected 3-pyridyl derivatives formed on combustion.

Alkaloids constitute approximately 2–4% by weight of fresh cigarette tobaccos (Kuhn 1965), with nicotine constituting roughly 95% of the total alkaloid fraction. The predominant secondary alkaloids, along with approximate percentages by weight of the alkaloid fraction, include: anatabine (2–3%), nornicotine (1%), anabasine (0.5%), and myosmine (0.1%) (Kuhn 1965; Leete 1983).

1.3 Pyrolysates of Tobacco Alkaloids

During smoking, approximately one-third of the nicotine present in tobacco is pyrolyzed, with the remaining two-thirds being distilled directly into the smoke (Kuhn 1965; Wynder and Hoffmann 1967). Accordingly, nicotine is expected to be (and is) the predominant tobacco alkaloid in smoke, including environmental tobacco smoke. However, the nicotine pyrolysis products play an increasingly important role in the chemical composition of sidestream and environmental tobacco smoke. To summarily classify these pyrolysates as alkaloids is open to debate (Kuhn 1965); however, we again take the wider view of *alkaloids* as espoused by Pelletier (1983) and supported by Leete (1983). Consequently, we consider all N-heterocyclic components

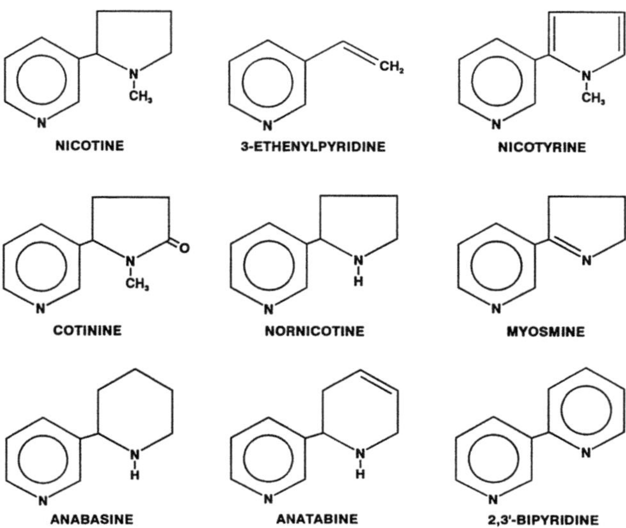

Fig. 1. Predominant alkaloids of tobacco and tobacco smoke

present in environmental tobacco smoke as the result of distillation and pyrolysis of the tobacco alkaloids to be the alkaloids of ETS.

The principal pyrolysis products of nicotine and related alkaloids are myosmine, nicotyrine, cotinine, bipyridine, and a series of simpler pyridine derivatives. Besides pyridine itself, the major derivatives are alkyl-substituted pyridines which include: 3-ethenylpyridine, 3-picoline (along with the 2- and 4- isomers), and the various isometric lutidines (Kuhn 1965). Structures of the most important tobacco and tobacco smoke alkaloids are illustrated in Fig. 1. Because of the extreme dilution of ETS in most indoor environments, many of these alkaloids are virtually nondetectable in real-world settings.

1.4 Tobacco-Related Alkaloids as Tracers of ETS in Indoor Air

One objective of our environmental tobacco smoke research program is to assess the impact of ETS on indoor air. Due to the chemical complexity, dynamics, and extreme dilution of ETS in indoor air, and the complexity of the indoor air background, it is impossible to ascertain the concentration level of ETS per se. Attempts are being made to put ETS concentrations into perspective by measuring selected ETS components which can then be used as markers or tracers for ETS as a whole. In this regard, any potential tracer should possess the following attributes: (1) it should be unique to or at least highly specific for tobacco smoke; (2) it should be in sufficient concentrations in air to be measured easily at realistic smoking rates; and (3) it should remain in constant proportion to the other components of ETS for a variety of tobacco blends and environmental conditions.

No single ETS component has yet been demonstrated to meet all these criteria successfully. Of all the known constituents of ETS, only the nicotine alkaloids (and selected pyrolysis products) and solanesol (Ogden and Maiolo 1989) appear to offer the required specificity.

Nicotine is by far the most commonly used indicator of ETS in indoor air. Many authors state erroneously that nicotine is unique to tobacco; however, such is not the case. Although present in a surprisingly large number of species other than *Nicotiana* (Leete 1983), and also detected at trace levels in a variety of common foods (Castro and Monji 1986; Sheen 1988; Davis et al. 1991), the presence of nicotine in indoor air should be uniquely attributable to tobacco smoke. The same should also hold true for the related alkaloids and the more unique pyrolysates (e.g., myosmine and 3-ethenylpyridine).

Nicotine may not be the ideal tracer, as once thought, due to unpredictable decay kinetics (Thome et al. 1986). Readily adsorbing to building materials and room furnishings, nicotine is depleted from ETS at a rate faster than most other components. As a result, some have suggested that nicotine concentrations underestimate concentrations of other ETS constituents (Eatough et al. 1990). This is undoubtedly true in many environments during the generation of smoke; however, the converse is true in environments with a recent past history of smoking. As demonstrated by Nelson et al. (1990), the adsorbed nicotine slowly desorbs over time and overestimates exposure to ETS in such environments. In these situations, ETS constituents other than nicotine decay to zero while nicotine decays rapidly to some standing background level with further decay occurring more slowly. Thus, measured concentrations of nicotine precisely assess only airborne nicotine and indicate only that smoking has taken place; they do not necessarily indicate the presence, and certainly not the levels, of other constituents.

Based on the unusual adsorptive and decay characteristics of nicotine, some investigators are turning their attention to 3-ethenylpyridine as a tracer of ETS (Thome et al. 1986; Eatough et al. 1989b; Ogden 1991). Second in concentration only to nicotine among the ETS alkaloids, 3-ethenylpyridine more nearly reflects the attributes of an ideal tracer discussed above than do any of the other tobacco alkaloids.

Several other of the minor secondary alkaloids and pyrolysates have received limited attention from one group of investigators (Benner et al. 1989; Eatough et al. 1989a,b, 1990). In addition to nicotine and 3-ethenylpyridine, these investigators also reported concentrations of myosmine, nicotyrine, cotinine, and 2-ethenylpyridine. With one exception (Eatough et al. 1989b), the only values reported were measured in a laboratory test chamber at greatly exaggerated smoke levels. Although of academic interest, these additional constituents do not at present figure predominantly in ETS research or indoor air quality investigations. An excellent review of the overall chemical composition of ETS has appeared recently (Guerin et al. 1992).

Due to sparsity of data in the literature, typical indoor air concentrations are difficult to summarize for all of the ETS alkaloids except nicotine. In a recent review of ETS nicotine measurements, Guerin et al. (1992, p. 151) compiled original data from 868 individual sampling locations and report nicotine concentrations $\leq 10\,\mu g/m^3$ in nearly 90% of the cases. Individual measurements can certainly exceed this value; however, levels greater than $100\,\mu g/m^3$ are rarely encountered.

Although 3-ethenylpyridine is considered by many to be a superior tracer of ETS vapor phase, there are few data in the literature from broad-based field surveys. In a 1988 survey of 41 restaurants (M.W. Ogden 1988, unpubl. data), the mean 3-ethenylpyridine concentration is $2.5\,\mu g/m^3$ (the corresponding nicotine mean is $10.5\,\mu g/m^3$). More recently (Ogden et al. 1993), the personal exposures of 96 nonsmoking women were monitored for a 5-day period resulting in a median 3-ethenylpyridine concentration of $0.15\,\mu g/m^3$ (the corresponding nicotine median is $0.43\,\mu g/m^3$).

Real-world data for other more ubiquitous alkyl-substituted pyridines are even more scarce. From data reported by Heavner et al. (1992) of a survey in eight homes, mean values are: 2-picoline $(0.2\,\mu g/m^3)$; 3-picoline $(0.3\,\mu g/m^3)$; 3-ethylpyridine $(0.1\,\mu g/m^3)$; and pyridine $(1.7\,\mu g/m^3)$.

Eatough et al. (1989b) report individual data from 20 indoor environments including homes, businesses, and discos. Mean values calculated from these data are: 3-ethenylpyridine $(3.2\,\mu g/m^3)$; 2-ethenylpyridine $(0.3\,\mu g/m^3)$; pyridine $(2.5\,\mu g/m^3)$; myosmine $(0.4\,\mu g/m^3)$; and cotinine $(0.3\,\mu g/m^3)$. The corresponding nicotine mean is $14.8\,\mu g/m^3$.

2 Overview of Sampling and Analysis Methods

Tobacco smoke is an aerosol, meaning that it is composed of vapor and particulate (tiny droplets suspended in the vapor) phases. Besides the physical difference between the aerosol components, there are chemical differences as well. The aerosol vapor phase is composed of the more volatile smoke constituents and many (if not most) of the semivolatile compounds that have evaporated from the particles of concentrated smoke. This phenomenon is illustrated best with nicotine, the major alkaloid component of tobacco smoke. A semivolatile compound, nicotine is believed to be in the particulate phase of both mainstream and sidestream smoke. However, due to evaporation, nicotine in ETS is found almost exclusively in the vapor phase (95+%) (Eatough et al. 1986; Eudy et al. 1986). Along this same line of reasoning, the more nonvolatile components of ETS reside in the particulate fraction of the aerosol. These differences dictate that sampling and analysis of the two aerosol phases are quite different. Guerin et al. (1992) present a detailed review of sampling and analysis strategies used in the measurement of ETS.

The most widely used techniques for sampling the ETS-related alkaloids include: sorbent bed sampling systems that collect vapor phase constituents on either XAD-4 resin for solvent extraction (Ogden 1989, 1991; Ogden et al. 1989a) or on Tenax for thermal desorption (Proctor et al. 1989; Thompson et al. 1989; Heavner et al. 1992); filter sampling systems that collect the more alkaline ETS alkaloids on acid-treated filters (Eatough et al. 1986, 1988; Hammond et al. 1987; Ogden and Maiolo 1992); diffusion denuder sampling systems (followed by acid-treated filters) for definitive separation of vapor and particulate phase alkaloids (Eatough et al. 1986, 1987, 1988, 1989b; Eudy et al. 1986); and diffusive sampling systems for collection of vapor phase nicotine (Eatough et al. 1987; Hammond and Leaderer 1987; Ogden et al. 1989b) or nicotine and 3-ethenylpyridine (Ogden and Maiolo 1992). Several of these sampling systems have been compared (Caka et al. 1990; Ogden and Maiolo 1992).

Of these methods, the XAD-4 sorbent collection with extraction in triethylamine-modified ethyl acetate (Ogden 1989; Ogden et al. 1989a) is the most widely documented, validated, and used method for nicotine determination. This methodology has been validated by ruggedness testing (Ogden 1989) and two international collaborative studies (Ogden 1989, 1992), and has been adopted as the official or recommended method by the following organizations: Association of Official Analytical Chemists (AOAC 1990), American Society for Testing and Materials (ASTM 1990), and the US Environmental Protection Agency (US EPA 1990).

Real-time monitoring of the ETS alkaloids has been performed with an atmospheric pressure chemical ionization mass spectrometer coupled to an environmentally controlled test chamber (Thome et al. 1986). Capable of determining airborne pyridine, ethenylpyridines, picolines, lutidines, nicotine, and ammonia in real-time at parts-per-trillion levels, this instrumentation was of paramount importance in the infancy of ETS research. Two discoveries, first made in 1985, have proven to be cornerstones in the present understanding of the role of nicotine in ETS. First presented in 1985 and published the following year (Eudy et al. 1986) was the discovery that ETS nicotine was virtually completely contained in the vapor fraction of the ETS aerosol and not with the particles (as had been assumed for many years from a knowledge of mainstream and sidestream smoke). The second was that the decay rate of nicotine was dramatically different from the other alkaloid constituents of ETS (Thome et al. 1986). While ethenylpyridine and pyridine decayed with nearly first-order kinetics, nicotine was shown to decay very rapidly initially and then to approach first-order decay at a considerable time after smoking. Based on these findings, these authors were the first to recognize that measured exposure to airborne nicotine (and even body burden of nicotine or cotinine) may have no quantifiable relationship to ETS exposure, as it may reflect exposure simply to nicotine and nothing else (Eudy et al. 1986).

3 Time-Integrated Sampling/Gas Chromatographic Analysis

3.1 Active Sampling and XAD-4 Sorbent Collection

This method has been validated for the determination of vapor-phase nicotine (Ogden 1989, 1992) and 3-ethenylpyridine (Nelson and Ogden 1990; Ogden 1991) in ambient air. Indications are that the methodology can be extended to the determination of vapor-phase myosmine; however, the methodology is not suited for the collection or determination of pyridine. Suitability for determining alkyl pyridines (picolines, lutidines, ethylpyridine, etc.) is not known.

3.1.1 Principle

A known volume of air is drawn at ≤ 1.5 l/min through a glass sampling tube containing XAD-4 resin. A minimum sampling time of 1 h is recommended. The XAD-4 resin is transferred to a GC autosampler vial and quinoline (internal standard) is added. Internal standard and collected analytes are extracted into solvent (ethyl acetate containing 0.01% v/v triethylamine) and an aliquot is analyzed by gas chromatography with thermionic-specific (i.e., N-selective) detection.

3.1.2 Apparatus

3.1.2.1 Pump

Samples are collected with a sampling pump calibrated for air flow rate of ≤ 1.5 l/min. [Note: sample flow rates >1.5 l/min result in breakthrough of 3-ethenylpyridine.]

3.1.2.2 Sorbent Sampling Tubes

Sample collection devices consist of glass sorbent tubes with both ends flame sealed, 7 cm × 6 mm o.d. and 4 mm i.d., containing two sections of 20/40 mesh XAD-4 resin (226-93 from SKC Inc., Eighty Four, Pennsylvania).

3.1.2.3 Gas Chromatograph

GC analyses are performed on a temperature-programmable system equipped with split/splitless capillary column injector, thermionic-specific (e.g.,

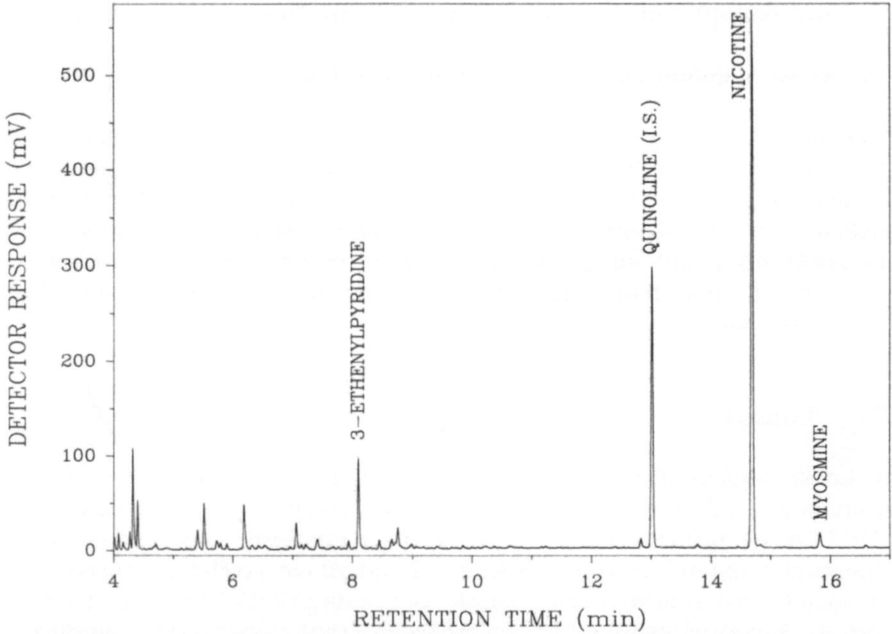

Fig. 2. GC/N-thermionic detection chromatogram of ETS vapor phase collected on XAD-4 resin

nitrogen-phosphorus) detector, digital electronic integrator, and autosampler (optional). Recommended capillary column is 30 m × 0.32 mm i.d. fused silica coated with a 1 μm film of 5% phenyl methylsilicone (DB-5, J&W Scientific, Inc., Folsom, California). Suitable operating conditions for the gas chromatograph are: helium carrier gas flow ca. 4 ml/min (15 psig); oven temperature 50 °C/hold 1 min, programmed at 10°/min to 250 °C; injector and detector temperatures 225 and 300 °C, respectively; injection volume 1–2 μl with 30-s splitless period. Suitable detector conditions are (or according to instrument manufacturer): helium make-up gas 15 ml/min; air 75 ml/min; hydrogen 3 ml/min; "bead current" sufficient to give S/N >50 for nicotine in 0.1 μg/ml calibration standard. A chromatogram of an ETS sample analyzed under these conditions is shown in Fig. 2.

3.1.2.4 Sample Containers

Samples are prepared in borosilicate glass autosampler vials (2 ml capacity) with Teflon-lined septum closures.

3.1.3 Reagents

The following reagents are necessary: ethyl acetate (chromatographic quality); nicotine (98+%, Eastman Kodak Co., Rochester, New York); 4-ethenylpyridine (95+%, Aldrich Chemical Co., Milwaukee, Wisconsin); quinoline (99 + %, Aldrich Chemical Co.); triethylamine (99+%, Aldrich Chemical Co.). [Note: 3-ethenylpyridine is not commercially available; instrument calibration is performed with the 4-ethenyl isomer according to Ogden (1991).]

3.1.4 Preparation of Standard Solutions

The ethyl acetate extraction solvent is prepared by adding triethylamine (0.01% v/v). Use this modified solvent (referred to hereafter as solvent) to prepare all standard and sample solutions.

To prepare primary standard solutions, weigh out separately 100 mg of each alkaloid (nicotine, 4-ethenylpyridine, etc.) into individual 100-ml volumetric flasks and dilute to volume with solvent. To prepare secondary standard solutions, pipet 1 ml of nicotine primary standard into 100-ml volumetric flask. Also, pipet 0.50 ml of 4-ethenylpyridine into the same volumetric flask (along with appropriate volume of any additional alkaloid primary solutions) and dilute with solvent.

Prepare quinoline (internal standard) solution by adding 10 µl quinoline to 100 ml solvent. [Note: quinoline can be detected in highly concentrated ETS at ca. 1% of the nicotine concentration (M.W. Ogden 1989, unpubl. data). This will not interfere with the quantitation of the remaining ETS alkaloids in the normal application of this method. However, if large amounts of ETS are concentrated, the standard procedure should be modified by either increasing the amount of quinoline internal standard added or by employing a different internal standard.]

Prepare at least two sets of five calibration solutions that cover concentration range of interest (typically 0.1–2 µg/ml nicotine) by adding, for example, 10, 20, 50, 100, and 200 µL of alkaloid secondary stock to each of five 2-ml glass vials. Add 50 µl quinoline secondary standard and 1 ml solvent to each vial, cap, and shake to mix. [Note: no volume corrections are necessary for standards prepared in this fashion because the weight ratios of analytes to quinoline are constant, regardless of additional solvent volumes added.]

3.1.5 Collection and Preparation of Samples

Break off both ends of the XAD-4 sorbent sampling tube. Position the backup section of XAD-4 resin nearest the pump and connect to the pump with tubing. Acquire samples at a flow rate of ≤1.5 l/min and record time elapsed during sampling. After sampling, cap the sorbent tube with plastic caps (supplied with tubes from manufacturer). Samples are stable for at least 2 weeks in the dark at room temperature; however, freezer storage is recommended. Stability of frozen samples is at least 4 weeks.

For analysis, transfer glass wool plug from the tube inlet and the primary section of XAD-4 resin to a 2-ml sample vial. Transfer backup section of XAD-4 resin along with bracketing glass wool plugs to a separate vial. [Note: if breakthrough information is not necessary from any particular sample, the entire tube contents may be transferred to a single vial.] Add 50 µl quinoline secondary standard solution directly onto the resin in each vial, add 1 ml solvent, cap vial and shake for 30 min (wrist-action shaker, Vortex mixer etc.).

3.1.6 Alkaloids Determination

Chromatograph one set of calibration standard solutions, samples, and second set of calibration standard solutions and obtain integrated areas for peaks of interest. Calculate analyte-to-quinoline peak area ratio for each. Average results for all calibration solutions analyzed at a given concentration and construct calibration curve for each analyte using peak area ratios as ordinate and micrograms analyte as abscissa values. Fit calibration data for each analyte to a regression model and calculate weight of analyte in each sample from analyte-to-quinoline peak area ratio. [Note: most thermionic-specific detectors exhibit nonlinearity in the concentration range specified in this method. Therefore, linear regression should be used with caution, especially at the lower concentrations specified. Superior results are usually obtained from a weighted $(1/x)$ linear or second-order least squares regression model.]

3.1.7 Desorption Efficiency Determination

Transfer the primary section of XAD-4 resin together with the associated glass wool plugs from 20 sample tubes to 20 2-ml vials. Spike five vials with 10-µl portions of alkaloid secondary standard solution, five vials with 20-µl portions, and five vials with 50-µl portions. The remaining five vials are blanks. Cap and store all vials in the same manner in which samples are handled. Uncap, add quinoline (internal standard) solution, desorb with solvent, and analyze all spiked samples and blanks as described above. Correct the results obtained on spiked samples by subtracting the blank average from the average for each spiked level. Calculate desorption efficiency (DE) for each alkaloid as follows:

DE = av. wt recovered (µg)/wt added (µg).

If DE is different for the three levels prepared (ANOVA or *t*-test, $P <$ 0.05), construct a plot of DE vs. weight of alkaloid found. If DE is equivalent for the three levels, but different from 100%, (*t*-test using pooled results, $P < 0.05$), use the pooled arithmetic mean for DE correction. DE should be determined two or more times until consistency is demonstrated.

3.1.8 Calculations

Correct the sample results for the blank by subtracting the average blank value (in µg) from the analyte weight determined on the front section of XAD-4 resin. Repeat for the backup section and add the two corrected values for a given sample to determine total weight analyte per sample (W_s). Correct the total weight of each analyte for the desorption efficiency determined for that analyte (if different from 100%; W_s/DE). The amount of each alkaloid determined in each XAD-4 sample tube is then converted to micrograms per cubic meter of air as follows:

$$\text{Analyte, µg/m}^3 = [\text{analyte mass (corrected) (µg)} \times 1000 \ (\text{l/m}^3)]/[\text{vol. air sampled (l)}],$$

where vol. air sampled (l) = pump flow rate (l/min) $\times$ sampling time (min).

If desired, results can be corrected to standard conditions of temperature and pressure (Ogden 1989); however, this is seldom done.

3.2 Active Sampling and Bisulfate-Treated Filter Collection

This method has been validated for the determination of total airborne nicotine; however, the methodology does not appear to be suitable for 3-ethenylpyridine collections (Ogden and Maiolo 1992). Likewise, this method is not applicable to the collection or determination of pyridine, the methyl pyridines (picolines), or other alkyl pyridines (lutidines, ethylpyridine, etc.). Indications are that the methodology can be extended to the determination of total airborne myosmine.

3.2.1 Principle

A known volume of air is drawn at ≤ 1.5 l/min through a filter cassette containing a sodium bisulfate-treated glass-fiber filter. A minimum sampling time of 1 h is recommended. The treated filter is transferred to a centrifuge tube and extracted in ethanol/water/sodium hydroxide solution and back-extracted into ammoniated heptane using N-ethylnornicotine as internal standard. An aliquot of the heptane layer is analyzed by gas chromatography with thermionic-specific detection.

3.2.2 Apparatus

3.2.2.1 Pump

Samples are collected with a sampling pump calibrated for air flow rate of
≤1.5 l/min. [Note: since this sampling system appears to be unsuitable for
collecting the more volatile, less alkaline alkaloids (e.g., 3-ethenylpyridine),
sample flow rates >1.5 l/min may be possible.]

3.2.2.2 Sodium Bisulfate-Treated Filters

Treated filters are prepared by saturating 37-mm glass-fiber filters
(TX40HI20WW, Pallflex Co., Putnam, Connecticut) in 4% sodium bisulfate
solution (prepared by dissolving 40 g sodium bisulfate monohydrate in 1 l of
distilled, deionized water). Filters are placed Teflon side up into a petri dish
containing the solution and allowed to stand ca. 30 s. Filters are then placed
onto a clean tray and dried in a vacuum desiccator for at least 1 h.

For sampling, filters are placed fiber side up into 37-mm cassette
housings (M000 037 00, Millipore Corp., Bedford, Massachusetts) on top of
a gasket (A31, Sloan Valve Co., Franklin Park, Illinois).

3.2.2.3 Gas Chromatograph

GC analyses are performed using the same equipment and conditions noted
for the XAD-4 procedure (see Sect. 3.1.2.3).

3.2.2.4 Sample Containers

Samples are prepared in disposable 15-ml polypropylene centrifuge tubes
with an aliquot of the final extract being transferred to borosilicate glass
autosampler vials (2 ml capacity) with Teflon-lined septum closures.

3.2.3 Reagents

In addition to the alkaloids listed in Section 3.1.3 (as necessary), the follow-
ing reagents are required: sodium bisulfate monohydrate (99%) and heptane
(99.9%) from Aldrich Chemical Co.; N-ethylnornicotine (NEN, Institut f.
Biopharmazeutical Mikroanalytik, Hamburg, Germany); absolute ethanol
(Quantum Chemical Corp., Tuscola, Illinois); 10 N sodium hydroxide (Fisher
Scientific, Fair Lawn, New Jersey); anhydrous, research grade ammonia
(Scott Specialty Gases, Plumsteadville, Pennsylvania); methanol (Burdick &

Jackson, Muskegon, Michigan); and high purity, distilled, deionized water (Barnstead NANOPure II water system, Barnstead Co., Dubuque, Iowa).

3.2.4 Preparation of Standard Solutions

To prepare primary standard solutions, weigh out separately 100 mg of each alkaloid (nicotine, etc.) into individual 100-ml volumetric flasks and dilute to volume with methanol. To prepare secondary standard solutions, pipet 4 ml of nicotine primary standard into 100-ml volumetric flask (along with appropriate volume of any additional alkaloid primary solutions) and dilute with methanol.

Prepare N-ethylnornicotine (internal standard) solution by adding 2.5 µl NEN to 100 ml ethanol. Ethanol is used as solvent to increase wettability of sample filters during extraction. [Note: NEN may coelute with myosmine under the chromatographic conditions specified. If attempting to determine myosmine, ensure adequate resolution is obtained and adjust GC parameters as necessary.]

Prepare at least two sets of five calibration solutions that cover concentration range of interest (typically 0.1–2 µg/ml nicotine) by adding, for example, 2.5–50 µl of alkaloid secondary stock to each of five 2-ml glass vials. Add 100 µl NEN stock and 1 ml ammoniated heptane to each vial, cap, and shake to mix. Heptane is ammoniated by bubbling ammonia for 2–3 min through a glass impinger containing heptane. [Note: methanol stock volumes greater than 50 µl should not be added to 1 ml heptane containing 100 µl ethanol. If higher concentration standards are needed, use a more concentrated stock solution.]

3.2.5 Collection and Preparation of Samples

Remove sealing plugs from filter cassette and position the back cassette half nearest the pump and connect to the pump with tubing. Acquire samples at a flow rate of ≤1.5 l/min and record time elapsed during sampling. After sampling, plug the cassette assembly with plastic plugs (supplied with cassette assemblies from manufacturer). Samples are stable for at least 2 weeks in the dark at room temperature; however, freezer storage is recommended. Stability of frozen samples is at least 4 weeks.

For analysis, transfer filter to a 15-ml centrifuge tube and add 100 µl internal standard solution directly onto the filter. Add 2 ml deionized water and mix tube contents vigorously (Vortex mixer) for ca. 15 s. Add 2 ml of 10 N sodium hydroxide and Vortex for ca. 15 s and add 2 ml ammoniated heptane and Vortex for 10 min (Big Vortexer, Glas-Col Apparatus Co., Terre Haute, Indiana). [Note: a single addition of 4 ml of 5 N sodium hydroxide instead of separate additions of water and 10 N sodium hydroxide has been found to be acceptable and time-saving.] Transfer a portion (ca. 1 ml) of the heptane layer (top) to a 2-ml glass vial for GC analysis.

3.2.6 Alkaloids Determination

Chromatograph standards and samples as described in Section 3.1.6. Likewise, obtain integrated peak areas and construct calibration curve from analyte-to-NEN peak area ratios.

3.2.7 Desorption Efficiency Determination

Load 20 cassette holders each with one treated filter. Spike five filters with 2.5-µl portions of alkaloid secondary standard solution, five filters with 5-µl portions, and five filters with 15-µl portions. The remaining five filters are blanks. Cap and store all cassettes in the same manner in which samples are handled. Transfer filters, add NEN (internal standard) solution, extract, and analyze all spiked samples and blanks as described above. Correct the results obtained on spiked samples for the blank average and calculate desorption efficiency (DE) as described in Section 3.1.7.

3.2.8 Calculations

Correct the sample results for the blank by subtracting the average blank value (in µg) from the analyte weight determined on each filter, correct for DE (if necessary), and calculate airborne concentration in µg/m^3 as described in Section 3.1.8.

3.3 Passive Sampling and Bisulfate-Treated Filter Collection

This method has been validated for the determination of vapor-phase nicotine and 3-ethenylpyridine in ambient air (Ogden and Maiolo 1992). Indications are that the methodology can be extended to the determination of vapor-phase myosmine. The methodology is not suited for the determination of pyridine. Suitability for determining alkyl pyridines (picolines, lutidines, ethylpyridine, etc.) is not known.

3.3.1 Principle

Volatile airborne alkaloids diffuse through a membrane filter windscreen and collect on a sodium bisulfate-treated glass-fiber filter. A minimum sampling time of 8 h is recommended. The treated filter is transferred to a centrifuge tube and extracted in ethanol/water/sodium hydroxide solution and back-extracted into ammoniated heptane using N-ethylnornicotine as internal standard. An aliquot of the heptane layer is analyzed by gas chromatography with thermionic-specific detection.

3.3.2 Apparatus

3.3.2.1 Passive Sampling Devices (PSDs)

Passive sampler housings are constructed from three-piece 37-mm filter cassettes (M000 037 A0, Millipore Corp., Bedford, Massachusetts). Windscreens for use in single-face exposure mode are made by cutting the larger diameter monitor ring section (i.e., the middle spacer) even with the inner lip and solvent welding a Schleicher & Schuell (Keene, New Hampshire) 37-mm TE39 polytetrafluorethylene filter (15-μm pore size) onto the cut face. Solvent welding is performed by saturating the windscreen filter in dichloromethane and pressing into place. Additional details of PSD construction, including clip attachment for personal monitoring, the construction of dual-face samplers, etc., are described in detail elsewhere (Ogden and Maiolo 1992).

3.3.2.2 Sodium Bisulfate-Treated Filters

Filters for use as collection media are saturated with sodium bisulfate as described in Section 3.2.2.2.

3.3.2.3 Gas Chromatograph

GC analyses are performed as described for the XAD-4 collection procedure in Section 3.1.2.3. A chromatogram of an ETS sample collected and analyzed according to this method is shown in Fig. 3.

3.3.2.4 Sample Containers

Samples are prepared in containers described in Section 3.2.2.4. Prior to, and immediately following sampling, the PSDs are stored in 4-ounce polypropylene specimen jars with screw-cap closures.

3.3.3 Reagents

The necessary reagents are those listed in Section 3.2.3 (including 4-ethenylpyridine as described in Sect. 3.1.3).

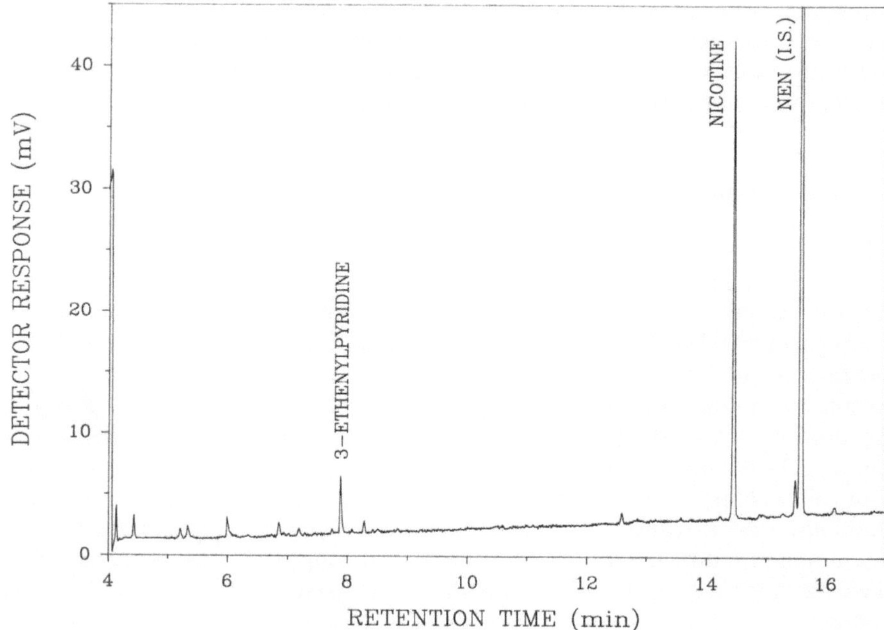

Fig. 3. GC/N-thermionic detection chromatogram of ETS vapor phase collected by diffusive (passive) sampling on a sodium bisulfate-treated glass fiber filter (chromatogram is qualitatively the same as would be obtained with active sampling and collection on a bisulfate-treated filter)

3.3.4 Preparation of Standard Solutions

Calibration standard solutions are prepared as described in Section 3.2.4.

3.3.5 Collection and Preparation of Samples

Prior to sampling, treated filters are placed fiber side up into the cassette bottom half and the windscreen assembly is pressed into place. No gasket or support pad is needed under the treated filter. Loaded samplers are stored in sealed specimen jars until use.

For sampling, remove PSD from polypropylene jar and position in area to be sampled (e.g., table top for area monitoring, attached to clothing near breathing zone for personal monitoring) ensuring that the windscreen face is unobstructed. At the end of the exposure monitoring period, record time elapsed during PSD exposure and return the PSD to the specimen jar. Samplers should be stored in a freezer until analysis. If possible, the bisulfate-treated filter should be removed from the PSD housing immediately after sampling and stored frozen in a clean centrifuge tube. Due to slight increases in nicotine on storage (Ogden and Maiolo 1992), PSDs should never be stored at room temperature after exposure.

Samplers should be cleaned between uses by washing in Micro detergent (International Products Corp., Trenton, New Jersey) followed by water and methanol rinses and air drying.

3.3.6 Alkaloids Determination

Chromatograph standards and samples as described in Section 3.2.6.

3.3.7 Desorption Efficiency Determination

Load 20 PSDs each with one treated filter and determine desorption efficiency as described in Section 3.2.7. Spiked filters may be stored either in the PSD housings (inside the specimen jars) or in clean centrifuge tubes. Conditions should be chosen which most resemble actual storage conditions of samples.

3.3.8 Calculations

Correct sample results for any blank and correct for DE (if necessary) as described in Sections 3.1.8 and 3.2.8. Conversion from analyte mass collected to air concentration is dependent on the uptake rate of the PSD for each analyte being determined. This PSD uptake rate is dependent on the physical dimensions of the sampler housing and the diffusion coefficient of the analyte. For the samplers described, uptake rates on single-face exposure are 31.5 ml/min for nicotine and 27.8 ml/min for ethenylpyridine (Ogden and Maiolo 1992). For maximum accuracy, uptake rates for additional alkaloids and for the actual samplers in use should be verified experimentally. In lieu of this, estimated sampling rates can be calculated (Ogden and Maiolo 1992). Calculate airborne concentration in $\mu g/m^3$ as described in Section 3.1.8, substituting PSD uptake rates for the pump flow rate.

4 Real-Time Sampling/Mass Spectrometric Analysis

4.1 Atmospheric Pressure Chemical Ionization Mass Spectrometry

The atmospheric pressure chemical ionization (APCI) mass spectrometer is a versatile tool which can be used for the determination of ETS-derived alkaloids in air samples. The instrument can be used for either confirmatory analysis or real-time quantitative analysis. Conventional mass spectrometers require that an analyte be introduced into a vacuum before analysis is completed. In APCI mass spectrometry, an analyte reacts with ions generated in a discharge at ambient pressure. Analyte ions are then extracted from the discharge into the vacuum region of the mass spectrometer.

Numerous detailed descriptions of the APCI operating principles appear in the literature (Thomson et al. 1980; Sunner et al. 1988a,b; SCIEX 1989). A brief description of the processes taking place in the ion source of an APCI mass spectrometer operated in the positive ion mode is given here. Electrons are generated in a point-to-plane electronic discharge in the APCI ion source. These electrons interact with nitrogen and oxygen to form positively charged ions. These ions then undergo charge exchange reactions with water vapor in the ion source to produce H_3O^+. These ions then undergo further aggregation reactions with water vapor in the ion source:

$$H_3O^+ + H_2O \rightarrow H_3O^+(H_2O) \tag{1}$$

$$H_3O^+(H_2O)_n + H_2O \rightarrow H_3O^+(H_2O)_{n+1}. \tag{2}$$

At 20 °C and ca. 21% relative humidity, the majority of clusters contain between five and eight water molecules (Sunner et al. 1988b). The hydrated H_3O^+ clusters are the principal ionizing reagent in the APCI ion source, and they protonate molecules (M) with gas phase basicities higher than water by the following reaction:

$$H_3O^+(H_2O)_n + M \rightarrow MH^+(H_2O)_x + (n + 1 - x)H_2O. \tag{3}$$

As the analyte ion cluster is drawn into the mass spectrometer, it passes through a curtain of dry N_2 gas. The water clustered to the analyte is removed in collisions with the curtain gas. Protonated molecular ions of the analyte are the predominant species observed in the APCI mass spectrum of most compounds.

$$MH^+(H_2O)_n \xrightarrow{N_2} MH^+ + n(H_2O). \tag{4}$$

Many compounds, including acids, can act as gas phase bases in their reactions with gaseous $H_3O^+(H_2O)_n$ ion clusters. However, the method is particularly suitable for the detection of compounds with high gas phase basicities. Nitrogen-containing bases tend to be particularly reactive in the APCI ion source. As a result, APCI provides an excellent method for detecting and determining concentrations of gas phase alkaloids. It should be noted, however, that this method is not suitable for the detection of any alkaloids which may be associated with the particulate phase of ETS.

The vapor phase of ETS does not contain significant numbers or concentrations of alkaloids distilled directly from tobacco (other than nicotine). However, pyrolysis of tobacco alkaloids such as nicotine leads to the formation of many N-heterocyclic compounds which can be found in the gas phase

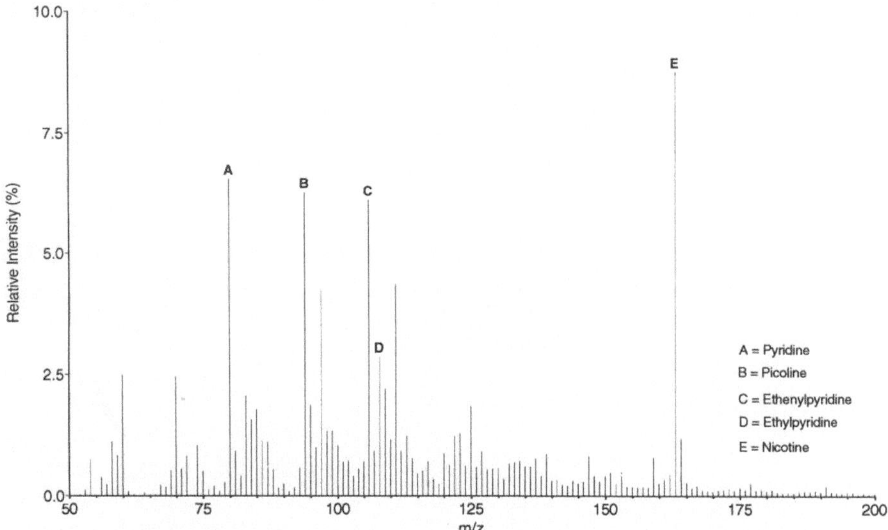

Fig. 4. Background subtracted APCI mass spectrum of environmental tobacco smoke

of ETS. Figure 4 shows a typical APCI mass spectrum obtained from ETS. The labeled peaks correspond to nicotine and four alkaloids derived from nicotine pyrolysis. Numerous other compounds are evident in the mass spectrum presented in Fig. 4, many of which are aliphatic amines and nitriles (such as propyl amine at m/z 60 and propionitrile at m/z 70) which are also readily detected by APCI/MS. The intense ion located at m/z 97 probably corresponds to the protonated molecular ion of an unidentified alkaloid with the molecular formula $C_5H_8N_2$ or $C_4H_4N_2O$. The ion appearing at m/z 111 probably corresponds to an analog of the compound at m/z 97 and differs by the addition of CH_2. Other tobacco-related alkaloids are presumed to be present in the APCI mass spectrum; however, the labeled alkaloids are the most abundant in ETS.

Since the pioneering work of Thome et al. (1986), many applications of APCI mass spectrometry to the detection and analysis of alkaloids and alkaloid-derived compounds in ETS have been developed. Qualitative analysis of alkaloids in ETS can be performed by APCI-MS/MS; however, this technique will not be discussed here. Real-time quantitative analysis is a highly useful technique for determining instantaneous compound concentrations and investigating the reactivity of the title compounds and their relationship and interactions to other compounds found in the indoor environment. Real-time data can be combined with plethysmography to accurately determine inhaled alkaloid dose (deBethizy et al. 1989). Analysis of the decay kinetics of ETS alkaloids can be used to understand relationships between various ETS tracers (Nelson et al. 1990, 1991). Time-weighted average concentrations of alkaloids in ETS can also be used to quantitatively describe differences between different cigarettes (Nelson et al. 1989).

4.2 Real-Time Quantitative Analysis

4.2.1 Calibration

To determine the concentrations of vapor phase alkaloids detected in ETS by APCI mass spectrometry, it is necessary to have vapor phase standards which can be used for instrument calibration. Gas dilution is perhaps the best way to calibrate for compounds in the gas phase. Gas dilution requires that a standard of known concentration and a method for accurately and reproducibly diluting the standard are available. Permeation tubes and diffusion tubes, housed in a constant temperature oven, are well suited for generating gas standards with known analyte concentrations. Table 1 includes the analyte, source, and typical source effusion rates used for investigating ETS along with the ion monitored for quantitative analysis of each analyte.

It is not possible to distinguish among different isomers of the substituted pyridines listed in Table 1 by APCI-MS; 3-picoline and 3-ethylpyridine were chosen for calibration because they are the predominant isomers of these compounds found in ETS. Although 3-ethenylpyridine is the most abundant ethenylpyridine isomer in ETS (there is a trace of the 2-ethenyl isomer and none of the 4-ethenyl isomer; Ogden 1991), it is not commercially available. 4-Ethenylpyridine is used for calibration because it is the most stable of the two commercially available isomers.

To generate a calibration curve for the determination of response factors, it is necessary to supply standards to the APCI ion source at different concentrations. The effluent from an oven containing permeation tubes is frequently too concentrated to be an effective calibration source. Furthermore, only one concentration is provided by a permeation tube in a constant temperature oven. To dilute the standard to working concentrations, a capillary dilution apparatus such as the one illustrated in Fig. 5 is used. The effluent from the oven is directed into the base of the apparatus. From here, the effluent can either pass through the capillary into a stream of dilution air

Table 1. Gas standard source type, source effusion rate, and m/z of ion monitored for alkaloids quantitated in environmental tobacco smoke

Compound	Source	Effusion rate (ng/min)	m/z Monitored
Nicotine	Diffusion tube	6000	163
Pyridine	Permeation tube	1000	80
3-Picoline	Permeation tube	550	95
4-Ethenylpyridine	Permeation tube	500	106
3-Ethylpyridine	Permeation tube	300	108

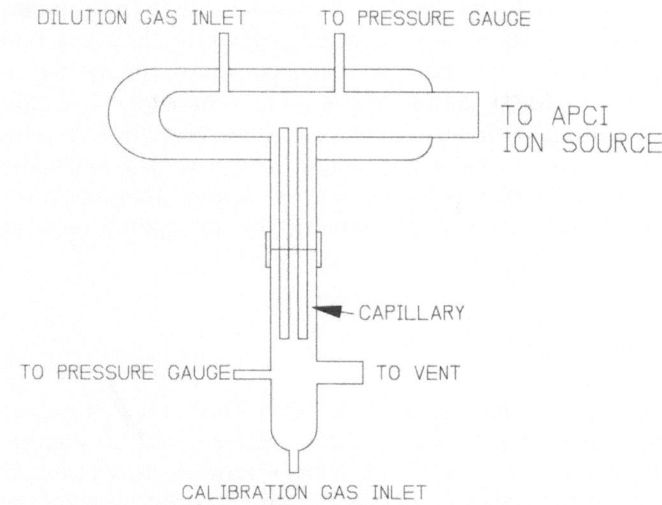

Fig. 5. Capillary dilution apparatus for preparing gas standards

(typically 8.7 l/min controlled by an electronic mass flow controller) or through a regulating valve leading to an exhaust. The flow of gas standard through the capillary can be adjusted by changing and controlling the pressure drop across the regulating valve. (In principle, this apparatus acts much like a split injector in a GC.) Pressure drop across the capillary is measured with a transducer. Analyte flow through the capillary is related experimentally to the pressure drop by measuring the flow with a bubble flowmeter. Once the relationship has been established, only sporadic flow checks are necessary to ensure that no leaks or changes in the system have taken place. By adjusting the valve to obtain known pressure drops/flows, a known volume of gas standard can be reliably and reproducibly blended with the dilution air stream. The concentration of analyte entering the APCI ion source can be determined easily from the volume and concentration of analyte passing through the capillary and the dilution air flow. The protonated molecular ion is the predominant species generated in the APCI ion source, and response factors are typically determined by selected ion monitoring of this ion at a range of concentrations bracketing the anticipated alkaloid concentration in ETS. In some cases, interfering ions may be present in ETS. If a fragment ion unique to the alkaloid can be determined by MS/MS, then multiple reaction monitoring any be necessary for quantitative analysis. Instrument response at each analyte concentration is used to generate calibration curves. Response factors are then obtained by linear regression of response vs. concentration.

There are a number of important considerations when using this method of instrument calibration. First, the concentration of the analyte exiting the permeation oven should be high enough that the flows through the capillary are small relative to the dilution flow. Second, adequate time must be allowed before measurements are made at each concentration. Analytes such as nicotine may adsorb to the glassware used, and it may take 5–10 min for a steady-state concentration of such analytes to be achieved. Third, the flow of dilution air must be carefully controlled. The use of a mass flow controller to achieve a steady flow is recommended. Lastly, the concentration range used for calibration should approximate the anticipated range of analyte concentrations in ETS.

4.2.2 Alkaloids Determination

The APCI mass spectrometer (TAGA 6000, SCIEX, Thornhill, Ontario) is a highly sensitive instrument capable of detecting analytes such as nicotine at the parts-per-trillion level. At high ETS concentrations, two problems may occur. First, the analyte signal may be so large as to overload the instrument. The second, more likely, problem is that high concentrations of an analyte such as nicotine may interfere with the determination of other analytes in the ETS matrix. It may be necessary to dilute the air entering the mass spectrometer to reduce the incidence of these two problems.

A block diagram illustrating an ETS sampling system is shown in Fig. 6. This apparatus serves to both sample and dilute ETS without the use of a mechanical pump. At the heart of the system is an Air-Vac vacuum transducer (Air-Vac Engineering Co., Milford, Connecticut). Dilution air passes through an annular orifice in the transducer. A partial vacuum is induced in the sampling tube and ETS is drawn into the vacuum transducer through this tube. Once in the vacuum transducer, the ETS sample is mixed with the dilution air, then passed to the APCI ion source of the mass spectrometer. The actual dilution factor can be changed by restricting the flow through the ETS sampling arm of the system. Typical dilution factors can range from 1:1 to 5:1 or more. Typically, the concentration of nicotine in the ion source should not exceed $50\,\mu g/m^3$ during sampling to minimize changes in instrument sensitivity (Nelson and Ogden 1990).

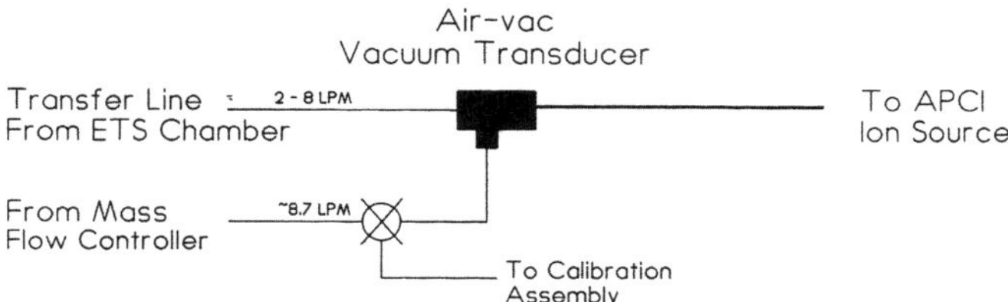

Fig. 6. Diagram of environmental tobacco smoke sampling system

In ETS, the protonated molecular ion of each analyte listed in Table 1 is normally the predominant ion appearing at that mass. The lack of interfering ions originating from other compounds occurring at the masses listed in Table 1 has been confirmed by MS/MS experiments. Real-time monitoring can be accomplished by selected ion monitoring of the relevant analyte ions in the ETS matrix over the course of an experimental period. Raw ion count data are then converted to concentrations using the instrument response factors. Excellent agreement between APCI mass spectrometry and the XAD-4 sorbent collection method described in Section 3.1 has been shown for ETS 3-ethenylpyridine (Nelson and Ogden 1990).

4.2.3 Internal Standard

The application of an internal standard to measurements made with an APCI mass spectrometer has been described (Nelson 1991). The principal components are a low concentration internal standard source and a flow delivery system for blending the internal standard with the air-stream entering the APCI ion source.

The choice of internal standard is critical to this system. The internal standard's ionization energetics and kinetics should be as similar as possible to those of the compound being quantified. Fully or partially deuterated analogs of an analyte normally should fulfill this condition. The m/z monitored for the internal standard should occur at a mass where no interfering ions appear. The presence of an internal standard must also not have a significant effect on the ionization efficiency of the analytes of interest. If ionization of the internal standard requires a large fraction of available reagent ions, then the sensitivity of the instrument to other analytes may be reduced. The internal standard is introduced into the sample stream as close as practical to the ion source. All components within the sample stream that come in contact with the internal standard are constructed from silanized glass to minimize reversible adsorption and unstable internal standard behavior.

The internal standard used for real-time nicotine determination in ETS is D_3(methyl)-nicotine. The ionization kinetics of this compound should be nearly identical to those of nicotine. The protonated molecular ion of this compound appears at m/z 166. Background ion intensity at this mass is insignificant with respect to the internal standard. D_3(methyl)-nicotine is not readily available in permeation tubes, and its vapor pressure at 30 °C is too great to be used as an internal standard. The internal standard was mixed with glycerol in a diffusion tube until it produced approximately 30 000 counts per second in the spectrometer. This corresponds to an internal standard evolution rate of ca. 80 ng/min and a fully diluted concentration of ca. 0.7 ppb. At these concentrations, the internal standard provides a signal large enough to be used for performing internal standard corrections, but not so large as to interfere with the sensitivity of the spectrometer to other ions of interest.

Internal standard data are collected along with real-time data, and may be used in several ways. If the internal standard concentration is known, then its response may be used to quantitate analyte response. Alternatively, the internal standard response may be normalized to its response at the beginning of the experiment in the absence of ETS. The normalized response may then be used to correct the analyte signal for changes in instrument sensitivity. Quantitation of the analyte would then proceed as normal, based on instrument response factors.

5 Summary

The major alkaloids in the vapor phase of environmental tobacco smoke are discussed and detailed procedures for their determination are outlined. Time-integrated sampling involves concentrating the alkaloids from the air matrix by trapping on either XAD-4 sorbent resin or on bisulfate-treated glass fiber filters with extraction and analysis by gas chromatography and N-selective detection. These methods are widely used in studying the impact of ETS on indoor air quality in real-life situations.

Additional methodology is described for real-time monitoring of the ETS alkaloids with an atmospheric pressure chemical ionization mass spectrometer. Capable of determining airborne alkaloids at parts-per-trillion levels, this method is invaluable for studying chemistry and kinetics at real-life concentration levels in a controlled laboratory setting.

Taken together, these methods form the core of our research program to study the use of alkaloids as tracers of environmental tobacco smoke and its impact on the chemical composition of indoor air.

References

AOAC (1990) Nicotine in environmental tobacco smoke: gas chromatographic method. In: Official methods of analysis, 15th edn, 1st suppl. Association of Official Analytical Chemists, Arlington, Virginia, method no 990.01

ASTM (1990) Standard test method for nicotine in indoor air. In: 1990 Annual book of ASTM standards. American Society for Testing and Materials, Philadelphia, Pennsylvania, vol 11.03, method no D 5075-90, pp 427–433

Benner CL, Bayona JM, Caka FM, Tang H, Lewis L, Crawford J, Lamb JD, Lee ML, Lewis EA, Eatough DJ (1989) Chemical composition of environmental tobacco smoke. 2. Particulate-phase compounds. Environ Sci Technol 23:688–699

Caka FM, Eatough DJ, Lewis EA, Tang H, Hammond SK, Leaderer BP, Spengler JD, Fasano A, McCarthy J, Ogden MW, Lewtas J (1990) An intercomparison of sampling techniques for nicotine in indoor environments. Environ Sci Technol 24:1196–1203

Castro A, Monji N (1986) Dietary nicotine and its significance in studies on tobacco smoking. Biochem Arch 2:91–97

Davis RA, Stiles MF, deBethizy JD, Reynolds JH (1991) Dietary nicotine: a source of urinary cotinine. Food Chem Toxicol 29:821–827

deBethizy JD, Bates LE, Davis RA, Heavner DL, Nelson PR, Walker JC, Robinson JH (1989) Nicotine absorption in humans following exposure to environmental tobacco smoke generated from different types of cigarettes. In: Bieva CJ, Courtois Y, Govaerts M (eds) Present and future of indoor air quality. Elsevier, Amsterdam, pp 269–276

Dube MF, Green CR (1982) Methods of collection of smoke for analytical purposes. Recent Adv Tob Sci 8:42–102

Eatough DJ, Benner C, Mooney RL, Bartholomew D, Steiner DS, Hansen LD, Lamb JD, Lewis EA (1986) Gas and particle phase nicotine in environmental tobacco smoke. Proc 79th Annual Meet Air Pollution Control Assoc. Air Pollution Control Assoc, Pittsburgh, paper 86-68.5

Eatough DJ, Benner CL, Bayona JM, Caka FM, Tang H, Lewis L, Lamb JD, Lee ML, Lewis EA, Hansen LD (1987) Sampling for gas phase nicotine in environmental tobacco smoke with a diffusion denuder and a passive sampler. Proc 1987 EPA/ APCA Int Symp Measurement of toxic and related air pollutants. Air Pollution Control Assoc, Pittsburgh, pp 132–139

Eatough DJ, Wooley K, Tang H, Lewis EA, Hansen LD, Eatough NL, Ogden MW (1988) Sampling gaseous compounds in environmental tobacco smoke. Proc 1988 EPA/APCA Int Symp Measurement of toxic and related air pollutants. Air Pollution Control Assoc, Pittsburgh, pp 739–749

Eatough DJ, Benner CL, Bayona JM, Richards G, Lamb JD, Lee ML, Lewis EA, Hansen LD (1989a) Chemical composition of environmental tobacco smoke. 1. Gas-phase acids and bases. Environ Sci Technol 23:679–687

Eatough DJ, Benner CL, Tang H, Landon V, Richards G, Caka FM, Crawford J, Lewis EA, Hansen LD (1989b) The chemical composition of environmental tobacco smoke III. Identification of conservative tracers of environmental tobacco smoke. Environ Int 15:19–28

Eatough DJ, Hansen LD, Lewis EA (1990) The chemical characterization of environmental tobacco smoke. Environ Technol 11:1071–1085

Eudy LW, Thome FA, Heavner DL, Green CR, Ingebrethsen BJ (1986) Studies on the vapor-particulate phase distribution of environmental nicotine by selective trapping and detection methods. Proc 79th Annual Meet Air Pollution Control Assoc, Pittsburgh, paper 86-38.7

Guerin MR, Jenkins RA, Tomkins BA (1992) The chemistry of environmental tobacco smoke: composition and measurement. Lewis, Chelsea, Michigan

Hammond SK, Leaderer BP (1987) A diffusion monitor to measure exposure to passive smoking. Environ Sci Technol 21:494–497

Hammond SK, Leaderer BP, Roche AC, Schenker M (1987) Collection and analysis of nicotine as a marker for environmental tobacco smoke. Atmos Environ 21:457–462

Heavner DL, Ogden MW, Nelson PR (1992) Multisorbent thermal desorption gas chromatography/mass selective detection method for the determination of target volatile organic compounds in indoor air. Environ Sci Technol 26:1737–1746

Kuhn H (1965) Tobacco alkaloids and their pyrolysis products in the smoke. In: von Euler US (ed) Tobacco alkaloids and related compounds: proc fourth int symp held at the Wenner-Gren center, vol 4. Stockholm, February 1964. Macmillan, New York, pp 37–51

Leete E (1983) Biosynthesis and metabolism of the tobacco alkaloids. In: Pelletier SW (ed) Alkaloids: chemical and biological perspectives, vol 1. John Wiley, New York, pp 85–152

Nelson PR (1991) Internal standard correction of results obtained by atmospheric pressure chemical ionization mass spectrometry. J Am Soc Mass Spectrom 2: 427–431

Nelson PR, Ogden MW (1990) Measurement of ethenylpyridine in environmental tobacco smoke. In: Proc 38th ASMS Conf on Mass spectrometry and allied topics. American Society for Mass Spectrometry, East Lansing, Michigan, pp 677–678

Nelson PR, Heavner DL, Collie BB (1989) Characterization of the environmental tobacco smoke generated by different cigarettes. In: Bieva CJ, Courtois Y, Govaerts M (eds) Present and future of indoor air quality. Elsevier, Amsterdam, pp 277–282

Nelson PR, Heavner DL, Oldaker GB III (1990) Problems with the use of nicotine as a predictive environmental tobacco smoker marker. Proc 1990 EPA/A&WMA Int Symp Measurement of toxic and related air pollutants. Air and Waste Management Assoc, Pittsburgh, pp 550–555

Nelson PR, deBethizy JD, Davis RA, Oldaker GB III (1991) Where there's smoke . . . ? Biases in the use of nicotine and cotinine as environmental tobacco smoke biomarkers. Proc 1991 EPA/A&WMA Int Symp Measurement of toxic and related air pollutants. Air and Waste Management Assoc, Pittsburgh, pp 449–454

Ogden MW (1989) Gas chromatographic determination of nicotine in environmental tobacco smoke: collaborative study. J Assoc Off Anal Chem 72:1002–1006

Ogden MW (1991) Use of capillary chromatography in the analysis of environmental tobacco smoke. In: Jennings W, Nikelly JG (eds) Capillary chromatography – the applications. Hüthig, Heidelberg, pp 67–82

Ogden MW (1992) Equivalency of gas chromatographic conditions in determination of nicotine in environmental tobacco smoke: minicollaborative study. J Assoc Off Anal Chem 75:729–733

Ogden MW, Maiolo KC (1989) Collection and determination of solanesol as a tracer of environmental tobacco smoke in indoor air. Environ Sci Technol 23:1148–1154

Ogden MW, Maiolo KC (1992) Comparative evaluation of diffusive and active sampling systems for determining airborne nicotine and 3-ethenylpyridine. Environ Sci Technol 26:1226–1234

Ogden MW, Davis RA, Maiolo KC, Stiles MF, Heavner DL, Hege RB, Morgan WT (1993) Multiple measures of personal ETS exposure in a population-based survey of nonsmoking women in Columbus, Ohio. Proc 6th Int Conf on Indoor air quality and climate. Indoor Air '93, Helsinki, vol 1, pp 523–528

Ogden MW, Eudy LW, Heavner DL, Conrad FW Jr, Green CR (1989a) Improved gas chromatographic determination of nicotine in environmental tobacco smoke. Analyst 114:1005–1008

Ogden MW, Nystrom CW, Oldaker GB III, Conrad FW Jr (1989b) Evaluation of a personal passive sampling device for determining exposure to nicotine in environmental tobacco smoke. Proc 1989 EPA/A&WMA Int Symp Measurement of toxic and related air pollutants. Air and Waste Management Assoc, Pittsburgh, pp 552–558

Pailer M (1965) Chemistry of nicotine and related alkaloids (including biosynthetic aspects). In: von Euler US (ed) Tobacco alkaloids and related compounds: proc fourth int symp held at the Wenner-Gren center, vol 4. Stockholm, February 1964. Macmillan, New York, pp 15–36

Pelletier SW (1983) The nature and definition of an alkaloid. In: Pelletier SW (ed) Alkaloids: chemical and biological perspectives, vol 1. John Wiley, New York, pp 1–31

Proctor CJ, Warren ND, Bevan MAJ (1989) Measurements of environmental tobacco smoke in an air conditioned office building. Environ Tech Lett 10:1003–1018

SCIEX (1989) The API book. SCIEX division of MDS, Thornhill, Ontario

Sheen S (1988) Detection of nicotine in foods and plant materials. J Food Sci 53:1572–1573

Stedman RL (1968) The chemical composition of tobacco and tobacco smoke. Chem Rev 68(2):153–207

Sunner J, Ikonomou MG, Kebarle P (1988a) Sensitivity enhancements obtained at high temperatures in atmospheric pressure ionization mass spectrometry. Anal Chem 60:1308–1313

Sunner J, Nicol G, Kebarle P (1988b) Factors determining relative sensitivity of analytes in positive ion mode atmospheric pressure ionization mass spectrometry. Anal Chem 60:1300–1307

Thome FA, Heavner DL, Ingebrethsen BJ, Eudy LW, Green CR (1986) Environmental tobacco smoke monitoring with an atmospheric pressure chemical ionization mass spectrometer/mass spectrometer coupled to a test chamber. Proc 79th Annual Meet Air Pollution Control Assoc. Air Pollution Control Assoc, Pittsburgh, paper 86-37.6

Thompson CV, Jenkins RA, Higgins CE (1989) A thermal desorption method for the determination of nicotine in indoor environments. Environ Sci Technol 23:429–435

Thomson BA, Davidson WR, Lovett AM (1980) Applications of a versatile technique for trace analysis: atmospheric pressure negative chemical ionization. Environ Health Perspect 36:77–84

US EPA (1990) Determination of nicotine in indoor air using XAD-4 sorbent tubes. In: Compendium of methods for the determination of air pollutants in indoor air. US Environmental Protection Agency, Research Triangle Park, North Carolina, EPA/600/S4-90/010, chap 2, method no IP-2A

Wolf FA (1967) Tobacco production and processing. In: Wynder EL, Hoffmann D (eds) Tobacco and tobacco smoke. Academic Press, New York, pp 5–45

Wynder EL, Hoffmann D (1967) Certain constituents of tobacco products. In: Wynder EL, Hoffmann D (eds) Tobacco and tobacco smoke. Academic Press, New York, pp 317–501

Methods for Production of Alkaloids in Root Cultures and Analysis of Products

J.D. HAMILL and A.J. PARR

1 Introduction

Plant cell and tissue culture has been used for many years to study and influence alkaloid biosynthesis from a wide range of plant species grown in culture. Nondifferentiated cultures have often been used, with variable degrees of success regarding the establishment and maintenance of productive cultures (Fujita 1990; Wilson 1990). In many cases considerable efforts have been expended attempting to maintain the cultures in a productive state with limited success (e.g., Deus-Neumann and Zenk 1984). Differentiated cultures such as root organ cultures, on the other hand, usually have the advantage of consistency in productivity of alkaloids over long periods of time in culture, which is beneficial for biosynthetic studies, precursor feeding experiments, and also the isolation and characterization of enzymes and c-DNAs for genes involved in alkaloid biosynthesis. Though known for 50 years to have this biosynthetic capacity (Dawson 1942), root organ cultures from many species have often been perceived to be difficult to maintain in vitro. In addition, the addition of hormones to culture media, which is often necessary to encourage rapid root growth, can lead to the situation arising where the optimal medium for growth is different in composition from the optimal medium for alkaloid production, as is the case for root cultures of *Hyoscyamus* (Hashimoto et al. 1986). In recent years, the difficulties involved in maintaining productive root cultures in vitro has been overcome for many species by transforming tissues with T-DNA of *Agrobacterium rhizogenes*. This soil bacterium has developed the capacity to insert T-DNA containing genes which alter the hormone metabolism of a wide range of dicotyledonous species such that roots differentiate and grow axenically in vitro in hormone-free medium, after the bacteria is killed by antibiotics. Often, these roots grow very rapidly, with cell mass doubling times being similar to that obtained using nondifferentiated cultures, although slow growing root cultures or shoot proliferation rather than root growth has been reported for alkaloid-producing *Cinchona* and *Papaver* species, respectively, following transformation with *A. rhizogenes* (Hamill et al. 1989; Yoshimatsu and Shimomura 1992). It is becoming clear that the interaction of *A. rhizogenes* transfer DNA (T-DNA) genes with the plant genome is quite sophisticated, the genes being subject to regulation by auxin and possibly other phytohormone levels in vivo (Maurel et al. 1990; Capone

Modern Methods of Plant Analysis, Volume 15
Alkaloids (ed. by Linskens/Jackson)
© Springer-Verlag Berlin Heidelberg 1994

et al. 1991). As these genes have been characterized (Slightom et al. 1986; Spena et al. 1987; Capone et al. 1989), and the biochemical function of their products becomes clear (Estruch et al. 1991a,b), it is possible to envisage the development of synthetic strains of *Agrobacterium* containing a gene complement which can be optimized for a particular species (e.g., *Cinchona* or *Papaver*) so that fast-growing and fully differentiated roots can be grown in vitro. In addition, as DNA transfer is involved in the production of transformed roots, it has become feasible to consider the genetic manipulation of alkaloid biosynthetic pathways to alter the productivity of such cultures (Hamill et al. 1990). A more detailed consideration of these issues is described in Hamill and Rhodes (1993).

This chapter is concerned with describing methods involved in production and characterization of transformed root cultures from alkaloid-synthesizing dicotyledonous plant species. It should be noted that a large number of secondary metabolites which are not classed as alkaloids have been reported to be produced by transformed root cultures (reviewed by Hamill and Rhodes 1993). Thus the technology is apparently applicable to production of any secondary metabolite produced by roots of dicotyledonous plants. A list of alkaloids reported to be produced by transformed root cultures is presented in Table 1.

Table 1. Examples of alkaloids produced by transformed root cultures which have been reported in the recent literature

Class of alkaloid	Genus	Reference
Pyridine	*Nicotiana*	Hamill et al. (1986); Parr and Hamill (1987)
Tropane	*Duboisia*	Deno et al. (1987); Mano et al. (1989)
	Datura	Flores and Filner (1985); Payne et al. (1987); Christen et al. (1989); Robins et al. (1990)
	Hyoscyamus *Scopolia*	Mano et al. (1986); Nabeshima et al. (1986)
	Atropa	Kamada et al. (1986); Jung and Tepfer (1987); Sharp and Doran (1990)
	Calystegia	Goldmann et al. (1990)
	Nicandra	Parr (1992)
Piperidine	*Lobelia*	Yonemitsu et al. (1990)
Alkamides	*Echinacea*	Trypsteen et al. (1991)
Indole	*Catharanthus*	Parr et al. (1988); Toivonen et al. (1989)
	Cinchona	Hamill et al. (1989)
	Amsonia	Sauerwein et al. (1991)
β-Carbolines	*Peganum*	Berlin et al. (1990)
Pyrrolizidine[a]	*Senecio*	Hartmann and Toppel (1987); Toppel et al. (1987)

[a] Rapidly growing nontransformed root cultures were used in these cases.

2 Induction of Transformed Roots

2.1 Choice of Bacterial Strains

Commonly used *A. rhizogenes* strains are agropine or mannopine producing, although strains capable of producing other opines, such as cucumopine, have been used to transform plant species (Bercetche et al. 1987). Agropine strains, which are often used, contain two sections of T-DNA, of about 20 kb, termed T_L and T_R (T Left and T Right, respectively) (Fig. 1; Jouanin 1984). *Rol* genes and other ORFs (open reading frames) which are important in causing roots to differentiate and grow are located on the T_L T-DNA (Spena et al. 1987; Schmulling et al. 1988; Capone et al. 1989), while the T_R T-DNA contains auxin biosynthesis genes which are functionally and structurally similar to the auxin synthesis genes (TMS1 and TMS2) of *A. tumefaciens* (Offringa et al. 1986). Agropine synthase is also located on the

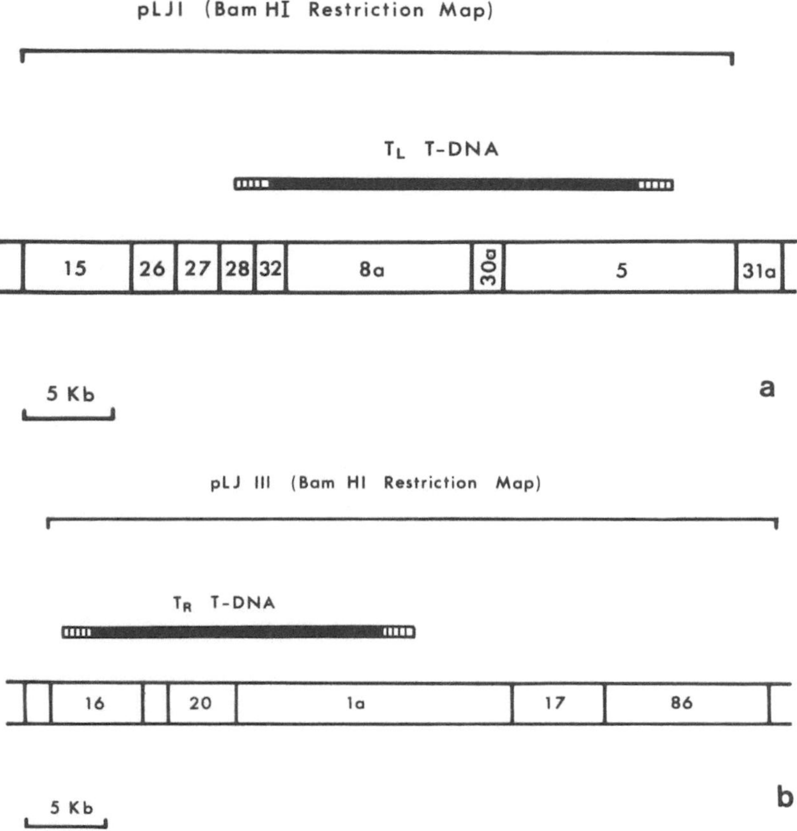

Fig. 1a,b. *Bam* HI restriction map of sections of Ri plasmid of agropine strains containing T-DNA. (After Jouanin 1984). **a** T_L T-DNA. **b** T_R T-DNA

T_R T-DNA (De Paolis et al. 1985; White et al. 1985). In our experience, agropine strains are very useful for inducing roots in a wide range of plant species and we have tended to favor the use of LBA9402 which contains the pRi1855 plasmid. This strain has been utilized by many groups and found to be infective on a wide range of dicotyledonous plants. Interestingly, it is also very effective in experiments involving DNA transfer to monocot plants using agroinfection (Boulton et al. 1989; Wilmink et al. 1992).

2.2 Culture of Bacteria

All *A. rhizogenes* strains grow well in YMB medium (Hooykaas 1988) consisting of 0.5 g/l K_2HPO_4, 0.2 g/l $MgSO_4$, 0.1 g/l NaCl, 0.4 g/l yeast extract, 10 g/l Mannitol (pH 7 before autoclaving).

For transformation, cultures are normally grown for 2 nights in liquid broth at 25 °C, with shaking, though overnight broths can be used if the colony used for inoculation is freshly grown from a fresh culture on an agar plate. *A. rhizogenes* will remain viable on YMB agar medium for long periods (6–12 months) if the plates are kept at 25 °C and are prevented from desiccation. For "permanent" storage, fresh 2-day-old broths can be diluted 1:1 with YMB containing 30% v/v glycerol (15% v/v final) and aliquots stored at −70 °C. Removal of frozen bacteria and inoculation onto modified YMB agar plates usually enables single colony recovery within 3–4 days.

2.3 Transformation of Plant Tissues

A wide range of plant tissues can be transformed with *A. rhizogenes* but we have found that inoculation of bacteria into leaf midribs or stem pieces tends to give the most consistent results. For medium-large leaves, such as tobacco, *Catharanthus*, etc., fully expanded leaves are removed from the upper part of the plant (preferably nonflowering) and surface sterilized by immersion in 10% v/v commercial bleach (e.g., Domestos) for 15–20 min followed by thorough rinsing with sterile distilled water. A similar protocol is used for surface sterilizing stem explants. Tissues damaged by the bleach are removed and *A. rhizogenes* is applied to the leaf midrib or stem explants as shown in Fig. 2A. Typically, tissues are wounded with a hypodermic needle and two to three drops (~20 µl) of bacterial suspension are applied to the wound spot (~10^{10} bacteria/ml). These leaves/stem pieces are then incubated in nutrient agar' (e.g., MS medium without phytohormones (Murashige and Skoog 1962), available from Sigma) or B5 medium without phytohormones (Gamborg et al. 1968, also available from Sigma) with 3% sucrose and 0.8% agar such that the infected portion is not in contact with the surface of the medium (Fig. 2B). When cultured at 25 °C, under illumination, e.g., 16 h day, 50 µE/m^2/s, roots usually differentiate around wound spots within 2

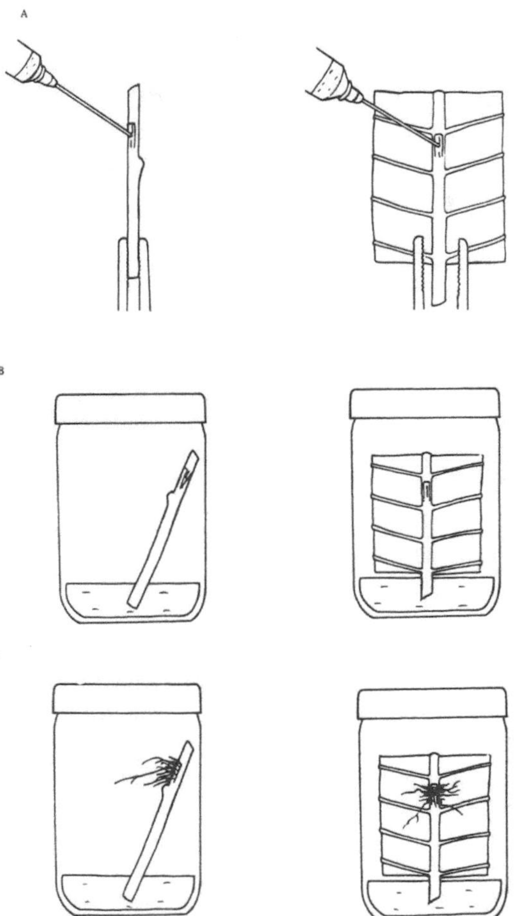

Fig. 2A–C. Diagrammatic representation of stem or leaf inoculation with *Agrobacterium rhizogenes* to produce transformed roots. **A** Surface-sterilized stem or leaf explants are placed on a sterile flat surface and gently wounded with a hypodermic needle. Two to three drops of *A. rhizogenes* suspension ($\sim 2 \times 10^8$ bacteria) are applied to the wound spot. **B** Tissues are cultured in nutrient agar (MS or B5 medium + 3% sucrose) such that the wounded portion does not come in contact with the medium. **C** After about 2 weeks at 20–25 °C, roots usually differentiate around the wound spot. These wound spots are cultured in 20–50 ml liquid medium, with 500 µg/ml cefotaxime or ampicillin to kill the bacteria

weeks (e.g., Fig. 2C, Fig. 3), though this can vary according to the plant species used.

For small leaf pieces or stem explants, or for tissues from tubers or other storage organs (e.g., carrot/potato slices), an approach similar to standard transformation procedures involving *A. tumefaciens* can also be

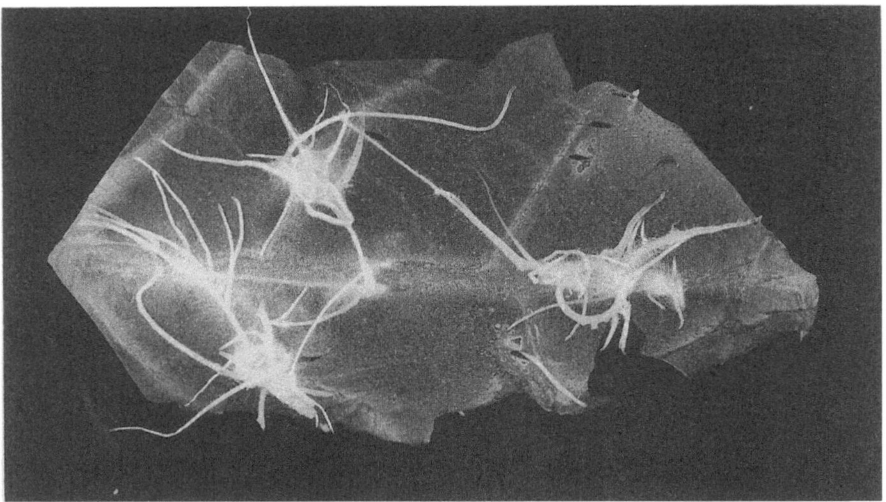

Fig. 3. Differentiation of roots on wounded leaf tissue of *Nicotiana tabacum* following inoculation with *A. rhizogenes* strain LBA9402

used. In these cases, a surface sterilized piece of tissue is dipped in a suspension of *A. rhizogenes* for 2 min, excess bacteria removed by placing on sterile blotting paper, and the tissue is cultured on sterile filter paper discs at 25 °C for 2 days. The choice of culture medium can vary. Tissues can be placed on sterile filter paper placed on a feeder layer of plant cells, e.g., tobacco cell suspension embedded in nutrient agar at a density about 10^4 cells/ml or, alternatively, on an agar/phytagel medium (such as MS or B5) containing phytohormones e.g., 0.5 mg/l BAP or 2 mg/l NAA and 0.5 mg/l BAP to stimulate cell division. After 2 days, tissues are rinsed thoroughly in sterile medium to remove excess bacteria and are then cultured on solidified MS or B5 medium with 3% sucrose containing 500 µg/ml ampicillin (sodium salt, Sigma) or 500 µg/ml cefotaxime (Claforan, Roussel) but devoid of exogenous phytohormones. (Antibiotics are added aseptically from a filter sterilized stock at 25 mg/ml, stored at −20 °C in small aliquots.) Tissues may need to be transferred to fresh medium containing antibiotics every 2–3 days if bacterial growth becomes evident. Under these conditions, roots usually differentiate within 10–14 days. Again, vascular tissue appears to be the most susceptible tissue for root production.

2.4 Culture of Root Tissues and Decontamination

In the majority of cases, culture of roots in hormone-free medium is very simple. A wound spot, containing many roots usually, is removed from the leaf/stem and placed in 20–50 ml liquid culture medium such as MS or B5 with 3% sucrose (pH 5.8 before autoclaving) and containing 500 µg/ml ampicillin or cefotaxime. In general, it is important not to remove the roots from the parent tissue until growth is evident. In some cases, e.g., *Catharanthus roseus*, a half- or quarter-strength solution of B5 medium containing 3% sucrose is beneficial to enable root cultures to become established. After 1–3 weeks, root growth is normally evident and a single root tip of about 2–3 cm in length is removed from the edge of the root mass and inoculated into a fresh batch of medium (with antibiotics) to enable establishment of a clonal culture. In the vast majority of cases, this approach ensures that fast-growing roots are selected for and it is unusual, in our experience, to detect nontransformed roots using this approach. An experiment to detect whether nontransformed roots do form under these conditions was carried out using *N. tabacum* and *B. vulgaris*. After roots had differentiated on leaf tissue, following transformation with *A. rhizogenes* LBA9402, individual roots were picked at random and removed from the explant. These were cultured in small volumes of medium until established. In the case of *B. vulgaris*, 11 root cultures were established, all of which were subsequently found to contain T_L and T_R T-DNA (Fig. 4A). In the case of *N. tabacum*, from ten roots placed in liquid medium, nine root cultures were established. Four of these lines did not contain any T-DNA and one of them contained only T_L T-DNA. The other four lines contained both T_L and T_R T-DNA (Fig. 4B). This experiment showed that nontransformed roots can differentiate around the wound spot in some species (such as tobacco) in addition to transformed roots, and that they may grow in culture without additional phytohormones. However, in the experiment noted above, nontransformed roots of tobacco were noticeably slower in growth than transformed roots. Thus, by following selection procedures noted earlier, i.e., allowing initial growth of all roots from wound spots followed by selection of a root tip from the edge of the colony, fast-growing transformed roots cultures will usually be the end result. However, it is advisable to check for the presence of T-DNA in all lines thought to be transformed, as described in Section 3.2.

2.5 Growth and Maintenance of Root Cultures

Root cultures grow best in liquid medium which is aerated by orbital motion of about 50 rpm. Usually, we grow them in 250-ml Erlenmeyer flasks or plastic culture vessels containing 50 ml liquid medium though they can grow in large containers. Growth in bioreactors, using a spray nozzle to deliver medium, is relatively straightforward and has been demonstrated for 20-l reactors. Conditions for growth in large bioreactors (500 l and above) are being determined (Wilson et al. 1990).

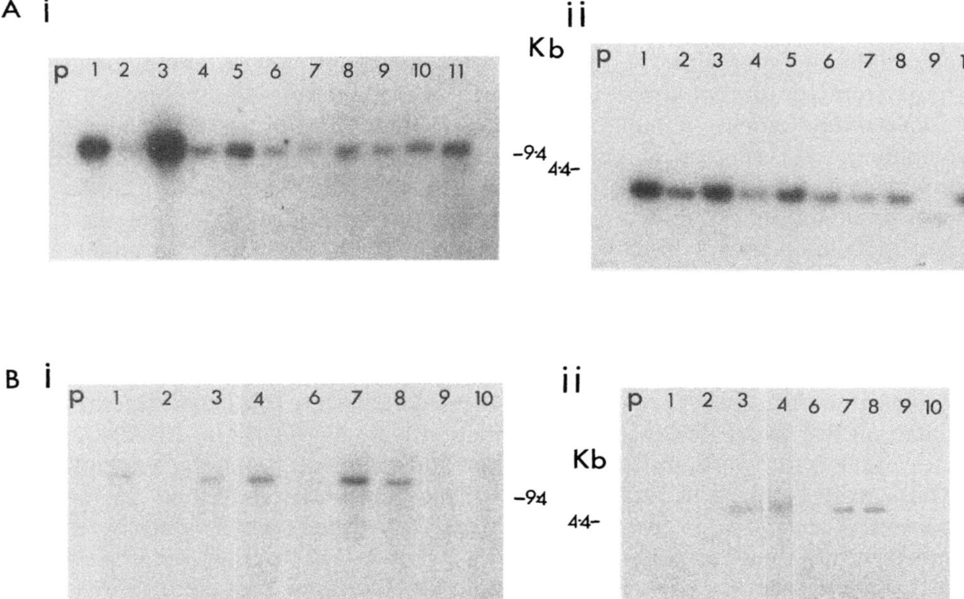

Fig. 4A,B. Detection of T_L and T_R T-DNA in root lines after establishment of cultures in vitro. DNA was extracted from 11 *B. vulgaris* root lines and 9 *N. rustica* root lines after establishment of cultures in vitro (line 5 of *N. rustica* was very slow-growing and did not yield enough DNA for analysis). Cultures were probed with ^{32}P-labeled *i* Bam HI 8a fragment from pLJ1 (T_L T-DNA); *ii* Bam HI 20 fragment from pLJIII (T_R T-DNA). **A** *B. vulgaris* root lines. **B** *N. rustica* root lines. All *B. vulgaris* root lines contained both T_L and T_R T-DNA (a truncated copy of T_R T-DNA was observed in *line 9*). Of nine *N. rustica* root lines analyzed, four contained T_L and T_R T-DNA (*lines 3, 4, 7,* and *8*), one contained only T_L (*line 1*), and the others (*2, 6, 9,* and *10*) did not contain T-DNA. These latter lines were noticeably slower in growth compared to those lines subsequently found to contain T-DNA

When grown in 250-ml Erlenmeyer flasks in 50 ml medium, the majority of root cultures require subculturing every 2–3 weeks, by which time a dense mat of tissue is often found. This is best done simply by aseptically removing a few roots from the edge of the culture and placing them in fresh medium. For medium-term storage, transformed roots will grow more slowly on the surface of solidified medium, e.g., at 8 g/l (Bacto-Difco) or Phytagel at 1.8 g/l (Sigma). At 25 °C, such cultures will usually require subculturing every 6–8 weeks (Fig. 5). Storage of tobacco transformed roots in liquid medium in gas permeable plastic containers at 4 °C is claimed to enable cultures to be maintained for periods in excess of 1 year (Tepfer, pers. comm.). For "permanent" storage of transformed root lines, it is necessary to consider storage in liquid nitrogen. This has been shown to be feasible for *Nicotiana rustica* and *Beta vulgaris* transformed root cultures with normal

Fig. 5. Transformed root culture of *N. rustica*, 6 weeks after subculture. The culture was grown on agar-solidified MS medium + 3% sucrose at 25 °C

growth rates and secondary metabolite production in lines after recovery from liquid nitrogen (Benson and Hamill 1991). Although the procedures need to be optimized before a generally applicable method can be reported, the protocol given below did enable about 20% of transformed roots of tobacco to survive freezing and produce normal transformed root cultures capable of synthesizing secondary products at pre-cryopreservation levels following thawing (Benson and Hamill 1991).

2.6 A Protocol for Cryopreservation of Transformed Roots
of *Nicotiana rustica*

Root tips of 2–4 mm in length are removed from the root culture at the end of the growth/early stationary phase. These are cryoprotected for 1 h at 25 °C in 0.5 liquid B5 medium with 3% w/v sucrose containing 10% U/V spectroscopic grade dimethylsulphoxide (DMSO) in 2 ml polypropylene cryovials (Sarstedt) at 25 °C. Vials are then cooled at 1 °C/min to a terminal temperature of 0 °C before being plunged into liquid N_2. For recovery, vials are rapidly thawed in a water bath at 45 °C for 2 min and roots placed on the surface of solidified B5 medium (5–10 root tips/9 cm dish). The presence of auxin (IAA or 2,4-D at 0.5 mg/l) in the medium was found to be beneficial for increasing the frequency of root redifferentiation and growth which, though variable, was about 20% on average (Benson and Hamill 1991). Root tips were cultured on solid medium for 2 weeks, after which time some callusing was evident in tobacco cultures. Tissues were then transferred to medium devoid of phytohormones, whereupon root redifferentiation and growth occurred within a few days. These roots were cultured in liquid B5 as for controls and showed growth rates and alkkaloid synthesis rates similar to nonfrozen parental tissues.

3 Proof of Transformation

For many workers whose primary interest lies in analysis of alkaloids produced by transformed roots, concrete proof of their transformed status is not required, and a fast growing and productive culture is sufficient. However, evidence of transformation is required in order to ascribe the term "transformed" to the root culture. This is easily carried out by probing a Southern Blot with T-DNA from *A. rhizogenes* or by carrying out a PCR reaction using oligonucleotides specific for one of the genes in the T-DNA. In addition, many authors report the use of an opine assay to detect T-DNA in the culture. Opine assays can, however, be an unreliable marker for transformation as it is possible for a root culture to contain T_L T-DNA and not T_R T-DNA (the location of agropine synthase genes in agropine strains), or for the opine synthesis genes to be present but not expressed. Thus, the analysis of opines is not discussed here (methods for their analysis have been published by Draper et al. 1988; Reynaerts et al. 1988).

3.1 DNA Extraction from Transformed Roots

Numerous protocols have been reported for extracting DNA from plants and the majority of these are quite suitable for extracting DNA from transformed roots. We routinely use a method described for leaf material by Martin et al. (1985) for isolating pure DNA. Crude "miniprep" DNA can be isolated by method of Lassner et al. (1989). Briefly these methods are as described in the following sections.

3.1.1 "Pure DNA" Isolation Procedure

A phenol solution is necessary and can be made as follows: To 500 g of high quality phenol, add 1.5 ml 10 N NaOH, 3 ml 2 M Tris HCl (pH 7.5) and 125 ml sterile distilled H_2O. Leave overnight at room temperature to dissolve. A phenol/chloroform solution is made by mixing phenol solution 1:1 with chloroform:isoamylalcohol 24:1. The phenol/ChCl$_3$ can be stored at 4 °C for 2–3 months in darkness and phenol/ChCl$_3$ solution is removed with minimal disturbance of the upper aqueous layer. Aliquots of phenol/ChCl$_3$ can be stored frozen at −20 °C for at least 2 years. To isolate DNA from transformed roots, excess moisture is removed with absorbent tissue from healthy root material grown in vitro and 5–10 g tissue is mixed with liquid N_2. The tissue is pulverized in an electric coffee grinder for about 30 s. It is important that tissue is ground to a fine dust while it remains frozen, or else the cell walls will not be efficiently broken to release sufficient DNA. The frozen powder is added to 10 ml extraction buffer (0.1 M sodium diethyl-thiocarbonate pH 7, 0.1 M Na_2 EDTA, 3 × SSC and 2% SDS) and incubated at 37 °C, with occasional shaking, until needed. Ten ml of phenol/chloroform is added and tubes are mixed thoroughly. Samples are centrifuged as before, supernatant is removed and extracted with 10 ml chloroform:isomylalcohol at 24:1. After centrifugation and re-extraction of the aqueous layer, the supernatant is removed and two volumes of 100% ethanol are added with mixing. At this stage, threads of DNA should be visible. DNA is pelleted by centrifugation and dissolved in 9 ml of TE (10 mM Tris HCl pH 7.5, 1 mM EDTA). Nine g of CsCl is added and dissolved and 1 ml ethidium bromide (10 mg/ml) is added. After placing in ultracentrifuge tubes and centrifuging at 40 K rpm for 40–60 h at 18 °C (60 000 rpm overnight), the fluorescing DNA band is visible under longwave UV illumination and is removed with a wide gauge needle. Ethidium bromide is removed by extracting two to three times with 5 ml of salt/H_2O saturated isoamylalcohol or isopropanol. The CsCl solution is diluted twofold with sterile distilled H_2O and DNA is recovered by the addition of 0.1 vol 3 M sodium acetate followed by 2 vol of 100% ethanol. Two cycles of dissolving in a small volume of distilled H_2O (e.g., 0.5 ml), followed by precipitation, usually produces DNA free of significant contaminants. DNA is finally dissolved in 100 µl TE and 5 ml used for DNA quantification (OD_{260} of 1.0 = 50 µg/ml of DNA). Good quality DNA has a ratio of approx 1:2:1 when read at OD_{230} (for carbohydrate detection), OD_{260} (DNA) and OD_{280} (for protein detection).

3.1.2 DNA "Miniprep" Method

The method of Lassner et al. (1989) for minipreparations of DNA works well for transformed roots. Extraction buffer consists of 0.14 M sorbitol, 0.22 M Tris-HCl (pH 8), 0.022 M EDTA, 0.8 M NaCl, 0.8% CTAB, 1.0%

N-laurylsarcosine and is made by mixing the CTAB (hexadecyl trimethyl ammonium bromide) and sarcosine together to form a paste before adding the rest of the solution. The EDTA is added last and the buffer is autoclaved.

To isolate DNA from transformed roots, take about 50 mg (freshly cut) of roots (about three roots of 2 cm length) and add to 500 µl of extraction buffer in a 1.5-ml centrifuge tube. Crush the tissue with a sterile glass rod or equivalent, ensuring that the buffer is well mixed. Add 300 µl of chloroform/ isoamylalcohol and top up with buffer to 1 ml. Invert the tube sharply several times. Incubate at 65 °C with lids open for 15 min, close lids, invert several times, and centrifuge at 14 000 rpm for 10 min. Remove 600 µl of upper aqueous phase carefully and transfer to a fresh tube containing 600 µl isopropanol. Invert the tube several times and centrifuge for 15 min. Pour off supernatant and wash the pellets carefully with 70% ethanol. Place in vacuum desiccator or speedy vac until almost dry and resuspend in 20 µl of TE; 2.5 µl of solution can be run on a minigel and stained with ethidium bromide to check that DNA has been extracted. Use a known quantity of marker DNA (e.g., nondigested phage λ) to estimate DNA concentration. The CTAB method can also be scaled up to isolate larger quantities of DNA with purification on a CsCl gradient if required.

3.2 Detection of T-DNA

3.2.1 Sources of Probes

The Ri plasmid is difficult to isolate due to its large size, and fortunately the plasmid from agropine strains has been isolated and the T-DNA subcloned by several groups (Laboratories of Dr. E. Nester, Seattle, USA; Dr. L. Jouanin, Versailles, France; Prof. P. Costantino, Rome, Italy). The pLJ series of cosmids are particularly useful, as together they encompass the entire Ri plasmid of strain HR1, which is almost identical with pRi1855 and pRiA$_4$b (Jouanin 1984). Cosmids pLJ1 and pLJ111 contain the entire T_L and T_R T-DNA, subcloned as partial *Bam*HI fragments, and give high yield of cosmid (1–2 mg from 500 ml of bacteria grown overnight) when conventional alkali lysis CsCl preparation methods are employed (Maniatis et al. 1982; Sambrook et al. 1989). A map of T_L and T_R T-DNA was shown in Fig. 1 (from Jouanin 1984). Digestion of transformed plant DNA with *Bam*HI enables T-DNA integration number and structural integrity of T_L and T_R T-DNA to be determined by judicial use of fragments to be used as a probe.

Digestion of pLJ1 with *Bam*HI releases, in addition to several smaller fragments, a large internal fragment of T_L T-DNA (*Bam*HI fragment 8a, 9.8 kb) and a fragment which spans the right T-DNA border/plant integration site of T_L T-DNA (*Bam*HI fragment 5, 12 kb).

Digestion of pLJ111 with *Bam*HI releases, in addition to several smaller fragments, an internal fragment (*Bam*HI fragment 20, 4.3 kb) a large fragment which spans the right T-DNA border/plant integration site of T_R T-

DNA (*Bam*HI fragment 1a, 16.8 kb) and also a fragment which spans the left T-DNA border/plant integration site of T_R T-DNA (*Bam*HI fragment 16, 5.5 kb).

In addition, *Bam*HI fragment 15 (from pLJ1, 6.1 kb) or *Bam*HI fragment 17 (from pLJ111, 5.2 kb) or 8b (from pLJ111, 9.8 kb) represent sequences which lie outside the T-DNA. These fragments can serve as internal control probes to ensure that a positive signal on an autoradiograph is caused by T-DNA integration into the plant genome rather than residual contamination of the sample by agrobacteria.

3.2.2 Recovery of DNA from Agarose Gels

A variety of approaches can be used to isolate DNA fragments from agarose gels to be used as probes (Sambrook et al. 1989). Larger pieces of DNA (10 kb and above) tend to be more difficult to purify using DEAE membranes, so yields are low. Commercially available kits such as Magic Minipreps (Promega) or Geneclean (Bio101) are alternative approaches for increasing yield. A simple method which works well for recovering DNA of all sizes from agarose gels is described below.

Separate fragments on a 1% gel using Tris borate buffer for electrophoresis (10 × TBE contains 108 g Tris base, 55 g boric acid, 9.3 Na_2 EDTA per l, pH 8.3. Filter and store in a dark glass bottle at room temperature). Stain gel with ethidium bromide (1 µg/ml in distilled H_2O) for 15 min, rinse, and identify bands of interest using a transilluminator. Cut out band using a sharp sterile scalpel and remove all excess agarose. Place the band (up to 1 cm in length) in a sterile 1.5-ml microcentrifuge tube and crush with a sterile glass rod. Add 500 µl of phenol solution and vortex to ensure that the agarose is fully suspended. Freeze at −70 °C for 1 h and then centrifuge at 14 000 rpm for 20 min. Remove supernatant, add 200 µl TE to phenol, vortex again, and recentrifuge again. Combine supernatants and extract with an equal volume of chloroform:isoamyl alcohol (24:1). Centrifuge at 14 000 rpm for 20 min, remove supernatant, add 0.1 vol 3 M sodium acetate and 2 vol of 100% ethanol. Place at −20 °C for 2 h/overnight. Centrifuge at 14 000 rpm for 30 min, remove supernatant carefully and wash DNA pellet with 50 µl 70% ethanol. Remove 70% ethanol and dry pellet carefully in vacuum dessicator/speedy vac and redissolve in 10 µl TE. Use 1 µl to run on a minigel in 1% agarose to check for successful isolation. Estimate concentration of band by running a known amount of standard (e.g., λ digested with *Hin*dIII or *Bst*EII).

3.3 Southern Blotting to Detect T-DNA in Tissues of Plant Material

Digest 10 µg of plant DNA from control (noninfected leaf tissue or root tissue) and putatively transformed root culture with *Bam*HI in a total volume of 30 µl. Add a total of 30–40 units of enzyme and incubate at 37 °C. Often several hours of digestion are required and the addition of spermidine (molecular biology grade, Sigma) to a final concentration of 1 mM can be helpful to aid digestion.

Check digestion is complete by running 2 µl on a TBE/1% agarose mini gel for 1 h at 80 mA and staining with ethidium bromide (1 µg/ml). If satisfied, run the remainder of the digest on a 1% agarose gel, using TBE as buffer, at 25–30 mA. Usually, it is desirable to run these gels overnight until the blue marker dye has migrated 15–20 cm down the gel.

After electrophoresis, the gel is stained with ethidium bromide and the gel photographed alongside a graduated ruler. It is desirable that marker bands of around 4 K have moved at least 10 cm down the gel to ensure that discrete bands will be seen when probing with T-DNA probes which hybridize across T-DNA border/plant junctions. Before transfer of DNA to membranes, the gel is treated with 0.25 M HCl for 20 min, then with 1.5 M NaCl/0.5 M NaOH for 40 min, and finally 1.5 M NaCl/1 M Tris HCl (pH 8) for 40 min. The gel is then blotted onto nitrocellulose or nylon membrane (e.g., Hybond C or Hybond N, Amersham), overnight using 20 × SSC as transfer buffer. (20 × SSC is 175.3 g NaCl/l, Na$_3$ citrate 88.2 g/l, pH 7 and autoclaved.)

The filter is washed gently with 100 ml 3 × SSC and air dried (DNA side up). For nitrocellulose, the membrane is baked at 80 °C (under vacuum) for 45 min–1 h (longer times make the membrane brittle). For Hybond N, the filter can be placed on the transilluminator for 30–60 s (this should ideally be calibrated for each transilluminator at regular intervals). If Hybond N$^+$ is used (Amersham), the DNA can be fixed for the membrane by treating with alkali followed by neutralization (15 min) and air drying the membrane. Other membranes can be used in conjunction with manufacturers' recommendations.

3.4 DNA Labeling and Hybridization

One hundred to 150 ng of probe DNA (as estimated by minigels) is ample to detect the presence of T-DNA in transformed root DNA when labeled with ^{32}P. We have tried ^{35}S and various nonradioactive labeling methods and have found them to be insufficiently sensitive to detect a single copy of a foreign gene when integrated into a relatively large genome such as tobacco. However, products which have recently become commercially available do claim to be as sensitive as ^{32}P and may be adequate for detecting T-DNA in DNA from transformed tissues. For ^{32}P labeling, commercially available kits are readily available which label DNA to high specific activity using the random primed method described by Feinberg and Vogelstein (1984). To make this buffer is also a relatively straightforward matter and requires three solutions.

3.4.1 Labeling Buffer

Solution I   625 µl of 2 M Tris HCl (pH 8); 25 µl 5 M MgCl$_2$; 5 µl each of 100 mM (pH 7) dATP, dTTP, dGTP (Pharmacia or Promega); 18 µl of 2-β-mercaptoethanol, 350 µl dH$_2$O.

Solution II   2 M HEPES titrated to pH 6.6 with NaOH, filter sterilized, and stored at 4 °C.

Solution III   Random hexanucleotides (e.g., Pharmacia Cat. No. 27 2166 01) resuspended in 3 mM Tris HCl, 0.2 mM EDTA (pH 7) at 90 OD$_{260}$ units/ml.

Combine solutions I, II, and III in the ratio 2:5:3 and freeze in aliquots of 50 µl. The resultant solution (designated labeling buffer) can be freeze thawed at least two or three times and is stable at −20 °C for at least 1 year.

3.4.2 DNA Labeling

To label DNA fragments, we typically use a labeling reaction of 50 µl. A "labeling mix" is made by combining 10 µl labeling buffer, 2 µl BSA (DNAase free, Pharmacia, 10 mg/ml, stored at 4 °C), 1 µl Klenow (5 u/µl) and is kept on ice. In a separate tube, combine DNA (50–150 ng) with sterile H$_2$O to give 33 µl and boil for 3–4 min. Pulse spin in microcentrifuge and place on ice for 2 min. Add boiled DNA solution to "labeling mix" and immediately add 5 µl ^{32}P dCTP (3000 Ci/mmol). Mix, pulse spin, and incubate at 37 °C for 2–3 h, or 25 °C for 4 h.

After incubation, add 50 µl TE buffer and purify probe from unincorporated nucleotides by passing through a Sephadex G50 (medium grade, Pharmacia) column equilibrated and autoclaved in TE buffer and poured in a sterile siliconized Pasteur pipet (Sambrook et al. 1989). Collect eight drop fractions and monitor elution profile using a Geiger counter. The labeled probe is usually eluted in fractions 3–5 and unincorporated nucleotides elute in tubes 6–8. Note: although some authors claim purification is not necessary, we have consistently found it is beneficial to purify the probe to reduce spots, blotches, and background of the subsequent autoradiograph.

3.5 DNA Hybridization

The DNA hybridization conditions may vary depending on the membrane used, and suppliers will often give details of optimal conditions to use with their membrane.

The following conditions work well with Hybond N and N$^+$ (Amersham) membranes using a prehybridization/hybridization buffer made according to the following recipe. For 50 ml of buffer, combine 40 ml sterile distilled water, 0.2 ml of 5% w/v Ficol (autoclaved and stored $-20\,^{\circ}\text{C}$), 0.2 ml 5% soluble polyvinylpyrrolidone (PVP) (autoclaved and stored $-20\,^{\circ}\text{C}$), 1 g of polyethylene glycol (PEG) (MW 8000, Sigma), 0.25 g skimmed milk powder, 3.8 ml SSPE [=1.5 $\times$ SSPE final, 20 $\times$ SSPE = 175 g NaCl, 31.2 g $NaH_2PO4.2H_2O$ made to 900 ml with distilled H_2O, adjusted to pH 7.7 with NaOH, 40 ml 0.5 M Na_2 EDTA (pH 8) added, made to 1 l, and autoclaved]. The above pre/hybridization solution is degassed for 5–10 min in a vacuum desiccator, and 3.5 ml 10% SDS and 2 ml of salmon sperm DNA which has been freshly boiled for 10 min are added. (Salmon sperm DNA is dissolved at 5 mg/ml in distilled H_2O, boiled for 10 min, sheared by passing through a 27-gauge needle three times and stored in 1 ml aliquots at $-20\,^{\circ}\text{C}$.)

Filters are prehybridized in the above solution for 2–3 h at 65 $^{\circ}\text{C}$ in a flat-bottomed plastic container (such as a heat-tolerant plastic lunch box) of the appropriate size. Sufficient liquid must be available to ensure that the membrane is just covered by the hybridization solution. The labeled probe is boiled for 10 min and added to this solution, and the membrane is incubated overnight with gentle shaking at 65 $^{\circ}\text{C}$. Wrapping the container in plastic cling film (e.g., Saran wrap) will ensure that evaporation is minimized. Filters are washed once with 2 $\times$ SSC/0.5% SDS and twice with 0.1% SSC/0.5% SDS for 30 min at 65 $^{\circ}\text{C}$ (high stringency) and while still damp, each filter is wrapped in cling film (e.g., Saran wrap) and exposed to X-ray film (e.g., Fuji RX) in a cassette with an intensifier screen at $-70\,^{\circ}\text{C}$. Normally, when probing for internal T-DNA sequences, these are visible with 1–2 days' exposure. T-DNA border/integration probes may require up to 1 week for clear signals to be seen.

3.6 Detection of T-DNA by Polymerase Chain Reaction (PCR)

This method is extremely rapid and is useful for determining whether a specific foreign gene is present in the DNA from a particular root culture. Full details are presented in Hamill et al. (1991). For *A. rhizogenes* T-DNA we have found primers specific to the *rol* B gene enables detection of the gene using both purified and crude DNA preparations from transformed roots. Oligonucleotides of 20–30 bases in length are suitable for amplification of foreign genes from genomic DNA by PCR. The specific sequences which are chosen will depend upon the precise sequence of the gene and thus it is important to check that any pair of primers will enable amplification of the appropriate gene using the *Agrobacterium* strain which was used as the source of the DNA. The PCR process is essentially the same using either plant DNA, or DNA from bacteria. Bacterial colonies can also be used without the need for DNA extraction if they are picked from fresh plates (2–3 days after streaking).

Many suppliers of Taq polymerase supply a concentrated buffer solution to which nucleotides, primers and DNA and enzyme need only be added. The final buffer typically contains 10 mM Tris HCl (pH 8.3 at 25 $^{\circ}\text{C}$) 1.5 mM

MgCl$_2$, 50 mM KCl, 0.01% gelatin (added from autoclaved sterile stock), 0.1% Triton X-100, 200 μM dATP, dCTP and dTTP, 100 μM dGTP, and 100 μM d^{7}GTP, 10 pmol of each primer (66 ng for a 20-mer) and 1 unit Taq polymerase. DNA concentrations of 20–200 ng are ideal. (This can be estimated for crude preparations.) Pre-boiling the plant DNA before adding to the reaction mix appears to assist in the reproducibility of the PCR reactions. It is advisable to assemble the reaction in 50 μl aliquots, heat the reaction to 94 °C, and cool to hybridization temperature before adding the Taq polymerase to ensure that no spurious synthesis of DNA occurs due to hybridization of primers to partially related sequences. In our experience, 25–30 cycles are adequate to see a band of predicted length when 20 μl of PCR reaction are run on a 1% agarose gel using TBE electrophoresis buffer. To confirm that the DNA band is from the gene of interest, it is necessary to carry out a Southern blot as before.

Oligonucleotides should be chosen such that the last two to three nucleotides at their 3′ ends are not complementary and ideally should have a similar GC%. The formula Tm = 2(A + T) + 4(G + C) is a reasonably accurate estimate of Tm for 20–21-mers and the formula Tm = 69.3 + 0.41(G + C)% − 650/N, where N = length of oligonucleotide, appears to give a reasonable estimate of Tm for oligos of 20–30 nucleotides in length (Hamill et al. 1991).

For amplification of the Rol B gene, the following oligonucleotides are suitable: 5′ATGGATCCCAAATTGCTATTCCTTCCACGA3′ and 5′TTAGGCTTCTTTCTTCAGGTTTACTGCAGC3′. These oligonucleotides allow the amplification of the *rol* B gene (ORF 11, 0.78 kb) coding sequence from transformed root tissues (Hamill et al. 1991). They are effective for detecting this gene in tissues transformed with a number of commonly used strains of *A. rhizogenes* including LBA9402 (agropine), NCCPPB2629 (agropine), 15834 (agropine), A4 (agropine), and 8196 (mannopine). It is most likely that shorter primers of 20–21 bases, from within these primers, would also be effective for detection of the *rol* B gene, though we have not tested this as yet.

Note: it is critical that adequate controls are done if PCR is used to detect specific sequences in tissues thought to be transformed. *A. rhizogenes* can persist, at low levels, in cultures, and the power of the technique is such that a few free-living bacteria could contribute to a positive result. It is important therefore to carry out control PCR reactions using oligonucleotides homologous to genes outside the T-DNA region. For agropine strains, we have found oligos from the vir D$_1$ gene of *A. rhizogenes* to be suitable for this purpose (Hamill et al. 1991). These are 5′ATGTCGCAAGGACGTAAGCCCA3′ and 5′GGAGTCTTTCAGCATGGAGCAA3′, which together enable a 0.45-kb fragment to be amplified by PCR.

Inverse PCR can, in theory, be used to estimate the number of insertions of T-DNA into a plant genome. A recent report by Does et al. (1991) described the use of this approach to detect the presence and number of T-

DNA/plant junction sites in transformed tobacco plants. Although described for binary vector-mediated transformation, the method could be used to determine the number of integrations of Ri T-DNA into transformed root lines.

4 Alkaloid Extraction and Quantification

Alkaloids have in the past been defined as organic bases which form water-soluble salts under acidic conditions, but which can be extracted into organic solvents under alkaline conditions. This operational definition provides the basis for a convenient extraction procedure. The roots are homogenized in dilute acid (typically $0.1\,M$ H_2SO_4, 5–10 ml per 1 g roots). After filtration through Miracloth to remove debris, the homogenate is extracted with organic solvent (1:1 vol), and the organic phase put aside. The aqueous phase is then made alkaline with NaOH or NH_4OH. Where possible, the pK of the alkaloid should be exceeded. The basified homogenate is then extracted twice with organic solvent. Chloroform and methylene dichloride are typically employed, but many others, e.g., ether, are also suitable. (Prolonged exposure of the alkaloids to chlorinated solvents should be avoided for very precise phytochemical work, since some chemical cross-reactions can occur.) The organic solvent is then removed under vacuum, and the alkaloid residue dissolved in a suitable solvent prior to HPLC or GC analysis (methanol can often be used; a modified version of the HPLC running buffer is also often advantageous). Some authors report a modification of the general procedure, so as to avoid working with large volumes of solvent. In these methods, the alkaline tissue homogenate is passed through a solid phase extraction cartridge (e.g., Extrelut, Merck), which binds the alkaloids. They are then eluted with a small volume of organic solvent. The alkaloid extract can sometimes be analyzed directly, or the solvent can be evaporated off and the residue redissolved as described above.

While these procedures work well, there are exceptions which need to be borne in mind. Certain alkaloids are unstable under alkaline conditions, and require modified extraction procedures. Tropane alkaloids are esters which undergo base catalyzed hydrolysis, but so long as the alkaloids are not exposed to a pH of above 11–12, and exposures are kept to less than 30 min, a conventional acid/base extraction regime can still be applied. Some alkaloids are quite hydrophilic even as bases (e.g., N-oxides, those with one or more free hydroxy groups, glycosides, etc.), and do not partition well into fully apolar solvents. In many cases, the use of a solvent such as ethyl acetate, or a $CHCl_3$-MeOH mixture (Phillipson and Handa 1975) will suffice. Sometimes, however, the use of another strategy may be required. This typically involves either omitting the final selective extraction step

Table 2. Typical examples of extraction protocols

Alkaloid class	Extraction method used	Reference
Nicotine	Acid/base partition	Hamill et al. (1986)
Tropane alkaloids	Acid/base partition	Payne et al. (1987)
Tropane alkaloids	Solid phase extraction	Mano et al. (1986)
Quinine	Acid/base partition	Robins et al. (1986)
Indole alkaloids	Acid/base partition	Renaudin (1984); Parr et al. (1988)
Calystegins	Ion exchange	Goldmann et al. (1990)
Glycoalkaloids	Solid phase extraction	Magrini et al. (1989)

altogether (e.g., Hartmann and Toppel 1987), or else carrying out a solid phase purification step by adsorbing the alkaloids onto a suitable column (perhaps a cation exchange resin, e.g., Goldmann et al. 1990) from which they are then selectively eluted by a carefully chosen solvent system. Table 2 shows typical examples of extraction protocols which have been used, though this list is by no means exhaustive.

Once a semi-purified alkaloid extract is obtained, the alkaloids require separation and quantitation. Baerheim Svendsen and Verpoorte (1983) have summarized an extensive literature on TLC analysis, but the methods of choice are now GC and HPLC (Verpoorte and Baerheim Svendsen (1983).) Several classes of alkaloid are sufficiently volatile to be analyzed in the underivatized state by GC (typically on a DB1 glass capillary column). This approach has a number of advantages; it is simple, allows the use of a specific nitrogen detector for selectivity, and by conversion to a GC/MS system, can give structural information on any unknowns present. Alkaloid classes that can be analyzed in this way include the *Nicotiana* alkaloids, tropanes, and many pyrrolizidines and quinolizidines. For many of the alkaloids found in roots, a HPLC method is, however, most appropriate. Although ion exchange and normal phase columns can be used, most published methods involve a C_{18} reversed phase column and a mobile phase containing a percentage of organic solvent such as methanol or acetonitrile. The precise composition of the solvent and the optimal pH will depend on the hydrophobicity and pKs of the alkaloids to be measured. Detection is typically by UV absorption, or by absorption in the visible region where the alkaloids are colored, e.g., berberine, serpentine, and sanguinarine. Certain alkaloids are also fluorescent (e.g., serpentine, β-carbolines), which provides a highly sensitive detection method. Not all alkaloids, however, contain a good chromophore, and here it is necessary to work at low UV wavelengths (200–210 nm). This is only possible with mobile phases that have little absorption in this region (using mineral acids rather than CH_3COOH), and sensitivity and selectivity will be low. A refractive index, rather than UV, detector can be also employed for alkaloids without a good chromophore, but this cases many of the same problems of selectivity and sensitivity. The high level of such alkaloids found in many root cultures (e.g., Fig. 6) mean,

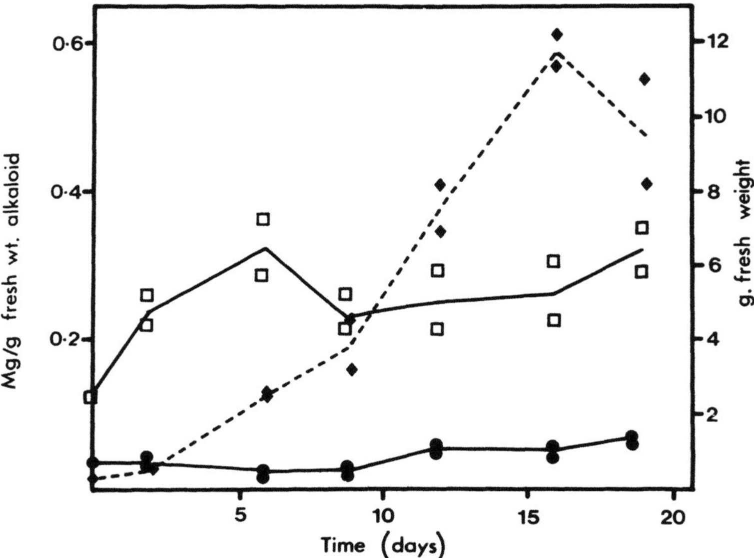

Fig. 6. Growth and alkaloid production by a transformed root culture of *Nicotiana rustica*. (Data redrawn from Hamill et al. 1986.) The culture was grown in liquid B5 medium + 3% sucrose at 25 °C. ♦–♦ fresh weight of tissue (blotted dry); □–□ nicotine (mg/g fresh wt); ●–● anatabine (mg/g fresh wt)

Table 3. Analytical methods for root-derived alkaloids

Alkaloid	Analysis method	Reference
Nicotine	HPLC (260 nm)	Robins et al. (1987)
Nicotine	GC	Wink and Witte (1987)
Tropanes	HPLC (230 nm)	Plank and Wagner (1986); Payne et al. (1987)
Tropanes	GC	Hartmann et al. (1986)
Quinine	HPLC (231 nm)	Robins et al. (1986)
Indole alkaloids	HPLC (254 nm)	Renaudin (1984)
β-Carbolines	HPLC	Berlin et al. (1990)
Pyrrolizidines	GC	Toppel et al. (1987)
Pyrrolizidine N-oxides	HPLC	Hartmann and Toppel (1987)
Calystegins	HPLC	Goldmann et al. (1990)
Bisbenzylisoquinolines	HPLC (282 nm)	Sugimoto et al. (1988)
Emetine	HPLC (280 nm)	Jha et al. (1988)
Solanum glycoalkaloids	HPLC	Magrini et al. (1989)

however, that in practice, such problems are usually not severe. A list of analytical methods used for a number of alkaloids is given in Table 3 and examples of HPLC condition for several alkaloids given in Table 4. Details on the use of HPLC in plant sciences, including analysis of alkaloids in

Table 4. Examples of HPLC systems for selected alkaloids (see references in text)

Alkaloid	Column	Solvent	Flow (ml/min)
Nicotine	C_{18}	$H_2O/CH_3CN/CH_3COOH/$"THF"[a] (430/12/3/1, pH 4.0)	1.2
Tropanes	C_{18}	$H_2O/CH_3CN/CH_3COOH/$"THF" (450/50/5/2)	1.0
Glycoalkaloids (*Solanum*)	C_{18}	0.01 M $(NH_4)_2HPO_4/CH_3CN/(C_2H_5)_3$ (67/33/0.02)	2.0
β-Carbolines	C_{18}	A = $H_2O/MeOH/HCOOH/(C_2H_5)_3$; B = MeOH (80/20/0.05,pH 8.5) Gradient 0% B-100% B over 15 min, then 100% B	1.0

[a] "THF" = tetrahydrofuran.

tobacco callus, were published recently in a previous book in this series (Linskens and Jackson 1987).

Acknowledgments. We are grateful to M. Sexton for the preparation of Fig. 2 and J. Elliston for preparation of the manuscript.

References

Baerheim Svendsen A, Verpoorte R (1983) Chromatography of alkaloids. Part A. Thin layer chromatography. J Chromatogr Lib, vol 23A. Elsevier, Amsterdam, 534 pp

Benson EE, Hamill JD (1991) Cryopreservation and post freeze molecular and biosynthtic stability in transformed roots of *Beta vulgaris* and *Nicotiana rustica*. Plant Cell Tissue Organ Cult 24:163–173

Bercetche J, Chriqui D, Adams S, David C (1987) Morphogenetic and cellular reorientations induced by *Agrobacterium rhizogenes* (strains 1855, 2659 and 8196) on carrot, pea and tobacco. Plant Sci 52:195–210

Berlin J, Mollenschott C, Greidziak N, Erdogan S, Kuzovkina I (1990) Affecting secondary product formation in suspension and hairy root cultures – a comparison. In: Nijkamp HJJ, van der Plas LHW, van Aartrijk J (eds) Progress in plant cellular and molecular biology. Kluwer Academic, Dordrecht, pp 763–768

Boulton MI, Buchholz WG, Marks MS, Markham PG, Davies JW (1989) Specificity of *Agrobacterium*-mediated delivery of maize streak virus DNA to members of the Graminaceae. Plant Mol Biol 12:31–40

Capone I, Spano L, Cardarelli M, Bellincampi D, Petit A, Costantino P (1989) Induction and growth properties of carrot roots with different complements of *Agrobacterium rhizogenes* T-DNA. Plant Mol Biol 13:43–52

Capone I, Cardarelli M, Mariotti D, Pomponi M, De Paolis A, Costantino P (1991) Different promoter regions control the level and tissue specificity of expression of *Agrobacterium rhizogenes rol* B gene in plants. Plant Mol Biol 16:427–436

Christen P, Roberts MF, Phillipson JD, Evans WC (1989) High yield production of tropane alkaloids by hairy-root cultures of a *Datura candida* hybrid. Plant Cell Rep 8:75–77

Dawson RF (1942) Nicotine synthesis in excised tobacco roots. Am J Bot 29:813–815
Deno H, Yamagata T, Emoto T, Yoshioka T, Yamada Y, Fijita Y (1987) Scopalamine production by root cultures of *Duboisia myoporoides*. II Establishment of a hairy root culture by infection with *Agrobacterium rhizogenes*. J Plant Physiol 131:315–323
De Paolis A, Mauro ML, Pomponi M, Cardarelli M, Spano L, Costantino P (1985) Localisation of agropine synthesizing functions in the T_R region of the root inducing plasmid of *Agrobacterium rhizogenes* 1855. Plasmid 13:1–7
Deus-Neumann B, Zenk MH (1984) Instability of indole alkaloid production in *Catharanthus roseus* cell suspension cultures. Planta Med 50:427–431
Does MP, Dekker BMM, de Groot MJA, Offringa R (1991) A quick method to estimate the T-DNA copy number in transgenic plants at an early stage after transformation, using inverse PCR. Plant Mol Biol 17:151–153
Draper J, Scott R, Armitage P, Walden R (1988) Plant genetic transformation and gene expression: a laboratory manual. Blackwell, Oxford
Estruch JJ, Chriqui D, Grossman K, Schell J, Spena A (1991a) The plant oncogene *rol* C is responsible for the release of cytokinins from glucoside-conjugates. EMBO J 10:2839–2895
Estruch JJ, Schell J, Spena A (1991b) The protein encoded by the *rol* B plant oncogene hydrolyses indole glucosides. EMBO J 10:3125–3128
Feinberg AP, Vogelstein B (1984) A technique for radiolabelling DNA restriction endonuclease fragments to high specific activity. Anal Biochem 137:266–267
Flores HE, Filner P (1985) Metabolic relationships of putrescine, GABA and alkaloids in cell and root cultures of Solanaceae. In: Neumann KH, Barz W, Reinhard E (eds) Primary and secondary metabolism of plant cell cultures, Springer, Berlin Heidelberg New York, pp 174–185
Fujita Y (1990) The production of industrial compounds. In: Bhojwani SS (ed) Plant tissue culture: applications and limitations. Elsevier, Amsterdam, pp 259–275
Gamborg OL, Miller RA, Ojima K (1986) Nutrient requirements of suspension cultures of soybean root cells. Exp Cell Res 50:151–158
Goldmann A, Milat M-L, Ducrot P-H, Lallemand J-Y, Maille M, Lepingle A, Charpin I, Tepfer D (1990) Tropane derivatives from *Calystegia sepium*. Phytochemistry 29:2125–2127
Hamill JD, Rhodes MJC (1993) Manipulating secondary metabolism in culture. In: Grierson (ed) Plant biotechnology III. Biosynthesis and manipulation of plant products. Blackie, London, pp 178–209
Hamill JD, Parr AJ, Robins RJ, Rhodes MJC (1986) Secondary product formation by cultures of *Beta vulgaris* and *Nicotiana rustica* transformed with *Agrobacterium rhizogenes*. Plant Cell Rep 5:111–114
Hamill JD, Robins RJ, Rhodes MJC (1989) Alkaloid production by transformed root cultures of *Cinchona ledgeriana*. Planta Med 55:354–357
Hamill JD, Robins RJ, Parr AJ, Evans DM, Furze JM, Rhodes MJC (1990) Overexpressing a yeast ornithine decarboxylase gene in transgenic roots of *Nicotiana rustica* can lead to enhanced nicotine accumulation. Plant Mol Biol 15:27–38
Hamill JD, Rounsley S, Spencer A, Todd G, Rhodes MJC (1991) The use of the polymerase chain reaction in plant transformation studies. Plant Cell Rep 10:221–224
Hartmann T, Toppel G (1987) Senecionine N-oxide, the primary product of pyrrolizidine alkaloid biosynthesis in root cultures of *Senecio vulgaris*. Phytochemistry 26:1639–1643
Hartmann T, Witte L, Oprach F, Toppel G (1986) Reinvestigation of the alkaloid composition of *Atropa belladonna* plants, root cultures and cell suspension cultures. Planta Med 52:390–395
Hashimoto T, Yukimure Y, Yamada Y (1986) Tropane alkaloid production in *Hyoscyamus* root cultures. J Plant Physiol 124:61–75
Hooykaas PJJ (1988) *Agrobacterium* molecular genetics. In: Gelvin SB, Schilperoort RA, Verma DP (eds) Plant Mol Biol Manual A4. Kluwer, Dordrecht, pp 1–3
Jha S, Sahu NP, Mahato SB (1988) Production of the alkaloids emetine and cephaeline in callus cultures of *Cephaelis ipecacuanha*. Planta Med 54:504–506

Jouanin L (1984) Restriction map of an agropine type Ri plasmid and its homologies with Ti plasmids. Plasmid 12:91–102

Jung G, Tepfer D (1987) Use of genetic transformation by the Ri T-DNA of *Agrobacterium rhizogenes* to stimulate biomass and tropane alkaloid production in *Atropa belladonna* and *Calystegia sepium* roots grown in vitro. Plant Sci 50:145–151

Kamada H, Okamura N, Satake M, Harada M, Shimimura K (1986) Alkaloid production by hairy root cultures in *Atropa belladonna*. Plant Cell Rep 5:239–242

Lassner MW, Peterson P, Yoder JI (1989) Simultaneous polymerase chain reaction amplification of multiple DNA fragments in the analysis of transgenic plants and their progeny. Plant Mol Biol Rep 7:116–128

Linskens HF, Jackson JF (1987) High performance liquid chromatography in plant sciences. Modern methods of plant analysis, vol 5. Springer, Berlin Heidelberg New York

Magrini E, Giulietti AM, Wilson E, Cascone O (1989) HPLC determination of glycoalkaloids in callus and fruits of *Solanum eleagnifolium*. Biotech Tech 3:185–188

Maniatis T, Fritsch EF, Sambrook J (1982) Molecular cloning: a laboratory manual. Cold Spring Harbor, New York

Mano Y, Nabeshima S, Matsui C, Ohkawa H (1986) Production of tropane alkaloids by hairy root cultures of *Scopolia japonica*. Agric Biol Chem 50:2715–2722

Mano Y, Ohkawa H, Yamada Y (1989) Production of tropane alkaloids by hairy root cultures of *Duboisia leichhardtii* transformed by *Agrobacterium rhizogenes*. Plant Sci 59:191–201

Martin CF, Carpenter R, Sommer H, Saedler H, Coen ES (1985) Molecular analysis of instability in flower pigmentation of *Antirrhinum majus*, following isolation of the pallida locus by transposon tagging. EMBO J 4:1625–1630

Maurel C, Brevet J, Barbier-Brygoo H, Guern J, Tempé J (1990) Auxin regulates the promoter of the root-inducing *rol* B gene of *Agrobacterium rhizogenes* in transgenic tobacco. Mol Gen Genet 223:58–64

Murashige T, Skoog F (1962) A revised medium for rapid growth and bioassays with tobacco tissue cultures. Physiol Plant 15:473–479

Nabeshima S, Mano Y, Ohkawa H (1986) Production of tropane alkaloids by hairy root cultures of *Scopolia japonica*. Symbiosis 2:11–18

Offringa IA, Melchers LS, Regensburg-Tuink AJG, Costantino P, Schilperoort RA, Hooykaas PJJ (1986) Complementation of *Agrobacterium tumefaciens* tumour-inducing aux mutants by genes from the T_R region of the Ri plasmid of *Agrobacterium rhizogenes*. Proc Natl Acad Sci USA 83:6935–6939

Parr AJ (1992) Alternative metabolic fates of hygrine in transformed root cultures of *Nicandra physaloides*. Plant Cell Rep 11:270–273

Parr AJ, Hamill JD (1987) Relationship between *Agrobacterium rhizogenes* transformed hairy roots and intact uninfected *Nicotiana* plants. Phytochemistry 26:3241–3245

Parr AJ, Peerless ACJ, Hamill JD, Walton NJ, Robins RJ, Rhodes MJC (1988) Alkaloid production by transformed root cultures of *Catharanthus roseus*. Plant Cell Rep 7:309–312

Payne J, Hamill JD, Robins RJ, Rhodes MJC (1987) Production of hyoscyamine by hairy root cultures of *Datura stramonium*. Planta Med 53:474–478

Phillipson JD, Handa SS (1975) N-Oxides of hyoscyamine and hyoscine in the Solanaceae. Phytochemistry 14:999–1003

Plank K-H, Wagner KG (1986) Determination of hyoscyamine and scopolamine in *Datura innoxia* plants by HPLC. Z Naturforsch 41C:391–395

Renaudin J-P (1984) Reverse-phase HPLC characteristics of indole alkaloids from cell suspension cultures of *Catharanthus roseus*. J Chromatogr 291:165–174

Reynaerts A, De Block M, Henalsteens JP, Van Montagu M (1988) Selectable and screenable markers. In: Gelvin SB, Schilperoort RA, Verma DP (eds) Plant Mol Biol Manual A9. Kluwer, Dordrecht, pp 1–16

Robins RJ, Payne J, Rhodes MJC (1986) Cell suspension cultures of *Cinchona ledgeriana*; I Growth and quinoline alkaloid production. Planta Med 52:220–226

Robins RJ, Hamill JD, Parr AJ, Smith K, Walton NJ, Rhodes MJC (1987) Potential for use of nicotinic acid as a selective agent for isolation of high nicotine-producing lines of *Nicotiana rustica* hairy root cultures. Plant Cell Rep 6:122–126

Robins RJ, Parr AJ, Walton NJ, Rhodes MJC (1990) Factors regulating tropane-alkaloid production in a transformed root culture of a *Datura candida* × *D. aurea* hybrid. Planta 181:414–422

Sambrook J, Fritsch EF, Maniatis T (1989) Molecular Cloning, 2nd edn. Cold Spring Harbor Laboratory Press, New York

Sauerwein M, Ishimaru K, Shimomura K (1991) Indole alkaloids in hairy roots of *Amsonia elliptica*. Phytochemistry 30:1153–1156

Schmülling T, Schell J, Spena A (1988) Single genes from *Agrobacterium rhizogenes* influence plant development. EMBO J 7:2621–2629

Sharp JM, Doran PM (1990) Characteristics of growth and tropane alkaloid synthesis in *Atropa belladonna* roots transformed by *Agrobacterium rhizogenes*. J Biotechnol 16:171–186

Slightom JL, Durand-Tardif M, Jouanin L, Tepfer D (1986) Nucleotide sequence analysis of TL-DNA of *Agrobacterium rhizogenes* agropine type plasmid. J Biol Chem 261: 108–121

Spena A, Schmülling T, Koncz C, Schell JS (1987) Independent and synergistic activity of *rol*A, B and C loci in stimulating abnormal growth in plants. EMBO J 6:3891–3899

Sugimoto Y, Sugimura Y, Yamada Y (1988) Production of bisbenzylisoquinoline alkaloids in cultured roots of *Stephania cepharantha*. Phytochemistry 27:1379–1381

Toivonen L, Balsevich J, Kurz WGW (1989) Indole alkaloid production by hairy root cultures of *Catharanthus roseus*. Plant Cell Tissue Organ Cult 18:79–93

Toppel G, Witte L, Riebesehl B, Borstel KV, Hartmann T (1987) Alkaloid patterns and biosynthetic capacity of root cultures from some pyrrolozidine alkaloid producing *Senecio* species. Plant Cell Rep 6:466–469

Tryptseen M, Van Lijsebettens M, Van Severen R, Van Montagu M (1991) *Agrobacterium rhizogenes*-mediated transformation of *Echinacea purpurea*. Plant Cell Rep 10:85–89

Verpoorte R, Baerheim Svendsen A (1983) Chromatography of alkaloids. Part B. J Chromatogr Lib, vol 23B. Elsevier, Amsterdam, 457 pp

White FF, Taylor BH, Huffman GA, Gordon MP, Nester EW (1985) Molecular and genetic analysis of the transferred DNA regions of the root-inducing plasmid of *Agrobacterium rhizogenes*. J Bacteriol 164:33–34

Wilmink A, Van de Ven BCE, Dors JJM (1992) Expression of the GUS gene in the monocot tulip after introduction by particle bombardment and *Agrobacterium*. Plant Cell Rep 11:76–80

Wilson G (1990) Screening and selection of cultured plant cells for increased yields of secondary metabolites. In: Dix PJ (ed) Plant cell line selection. VCH, Weinheim, pp 187–213

Wilson PDG, Hilton MG, Meehan PTH, Waspe CR, Rhodes MJC (1990) The cultivation of transformed roots from laboratory to pilot plant. In: Nijkamp HJJ, Van der Plas LHW, Van Aartrijk J (eds) Current plant science and biotechnology in agriculture: progress in plant cellular and molecular biology. Kluwer, Dordrecht, pp 700–705

Wink M, Witte L (1987) Alkaloids in stem roots of *Nicotiana tabacum* and *Spartium junceum* transformed by *Agrobacterium rhizogenes*. Z Naturforsch 42C:69–72

Yonemitsu H, Shimomura K, Satake M, Mochida S, Tanaka M, Endo T, Kaji A (1990) Lobeline production by hairy root culture of *Lobelia inflata L*. Plant Cell Rep 9:307–310

Yoshimatsu K, Shimomura K (1992) Transformation of opium poppy (*Papaver somniferum*) with *Agrobacterium rhizogenes* MAFF03-01724. Plant Cell Rep 11:132–136

Genetic and Chemical Analysis for Alkaloids in Papaver

J.R. Sharma and M.M. Gupta

1 Introduction

In the preface of the monograph *The Opium Poppy*, the editors Husain and Sharma (1983) have aptly summed up the significance of *Papaver*, *P. somniferum* in particular, in the growth of major world civilization. They stated: "The story of the man's civilization is, by and large, a chronicle of his socio-economic development based on the exploitation of enormous plant resources which nature has generously bestowed upon the earth. Though food plants were its pivotal base, other plants equally contributed to its grandeur. For instance, the growth of major world civilizations (Egyptian 1955 B.C., Greek 1500–1000 B.C., Mesopotamian 700 B.C.) is closely associated with recognition of the opium poppy as a useful plant of medicinal significance in pre-historic times."

Even today, notwithstanding the illicit production and illegal transaction of opium in the countries belonging to the so-called Golden Triangle (Thailand, Burma, and Laos) and Golden Crescent (Afganistan, Iran, and Pakistan), opium and its constituent alkaloids are of very great value in modern medicine, though their social abuses are potentially terrifying. The ultimate source of opium and opium alkaloids is axiomatically the opium poppy – *Papaver somniferum*, species of the genus *Papaver*. Among numerous alkaloids occurring in the *Papaver* species, the 40 alkaloids isolated from opium poppy (Šantavý 1970, 1979; Thakur 1983), could be grouped into nine groups or classes (each consisting of one or more alkaloids), viz., morphinane, benzylisoquinoline, aporphine, proto- and tetra-hydroberberine, protopine, phthalideisoquinoline, rhoeadine, benzophenanthridine, and tetrahydroisoquinoline. The scope of this chapter, however, does not permit us to give details about all these alkaloids; only the five major alkaloids, namely morphine, codeine, thebaine, papaverine, and narcotine, are elaborated in this chapter with particular respect to their chemical analysis.

However, before we embark upon their chemical assaying, two points merit consideration. Firstly, owing to the tremendous pharmacological and clinical significance of opium and opium alkaloids, two species of the genus *Papaver*, viz., *P. somniferum* (section *Papaver*) and *P. bracteatum* (section *Oxytona*) are under wide cultivation. Secondly, as a result of their being under cultivation, it is only logical to know a priori the genetics of the major opium alkaloids to be able to harness their maximum productivity per unit area, time, and input through evolving genetically superior strains. Therefore, in the following section, we intend: (1) to provide an over-view of the genetics of major opium alkaloids, and (2) to signify the associative role of their chemical analysis.

Modern Methods of Plant Analysis, Volume 15
Alkaloids (ed. by Linskens/Jackson)

2 Genetics of Major Opium Alkaloids and Importance of Chemical Analysis

Since Nyman and Hall (1976), 16 years ago, noted that: "very little of the inheritance of the opium alkaloids is known to-day, although varietal studies on this were started 35 years ago" (Asahina et al. 1957; Böhm 1967; Lalezari et al. 1974; Shafiee et al. 1975; cf. Sharma and Singh 1983, etc.), there has been considerable progress in the knowledge of gene control and dominance relationships of these alkaloids in the two sections *Papaver* (Mecones) and *Oxytona* (Macrantha) of the genus *Papaver*, given below.

2.1 Genetics of Alkaloids of Sections *Papaver and Oxytona*

The section *Papaver* comprises two major species: *P. somniferum* ($2n = 22$) and *P. setigerum* ($2n = 44$). Both are capable of synthesizing opium alkaloids in their latex, but the genetic assaying of major alkaloids has been carried out only in the former. Based on several experimental studies, all three alkaloids, morphine, codeine, and thebaine, were found to be influenced predominantly by additive genetic variance, although the nonadditive component was also significant for morphine content (Srivastava and Sharma 1987; Sharma et al. 1988; Lal and Sharma 1991). Narcotine content was, however, conditioned largely by the nonadditive component. The role of overdominance was great for all these alkaloids. Heritability estimates ranged from low to high for morphine, but were moderate for codeine and thebaine, and low for narcotine contents. Further, the reciprocal effect was considerable for morphine and moderate for narcotine, but negligible for codeine, and nonconsequential for thebaine content. Duplicate epistasis was very effective for morphine and codeine but not for thebaine and narcotine contents. No information is available on papaverine content.

Based on this information, maximum heterosis (hybrid vigour) can be registered for thebaine and narcotine, but only low heterosis for morphine and codeine alkaloids (Khanna and Shukla 1986; Lal 1988; Sharma et al. 1988). Moreover, owing to the high amount of fixable genetic variance associated with morphine, codeine, and thebaine, strain selection can also lead to rewarding results in *P. somniferum*.

However, increasing abuses of the opium obtained from *P. somniferum* have necessitated the search for new and cheaper alternate sources that can eliminate such abuses. The fact that thebaine is the natural precursor of codeine attracted attention to the section *Oxytona* (Shafiee et al. 1975; Fairbairn 1976; Nyman and Bruhn 1979; Šantavý 1979; Levy et al. 1981; Levy 1985; Milo et al. 1987, 1988, 1990; Chelombit'Ko and Mikheev 1988;

Laane et al. 1988; Levy and Milo 1991), which offers a natural source of thebaine. The section *Oxytona* includes the three species: *P. bracteatum* (PB) (2n = 14, diploid), *P. orientale* (PO) (2n = 28, tetraploid), and *P. pseudo-orientale*) (PPO) (2n = 42, hexaploid) (Goldblatt 1974; Theuns et al. 1987).

The thebaine content in PB ranges from 0.82 to 3.8% in capsule husk and 0.45 to 1.23% in the roots. In fact, thebaine constitutes about 98% of the total alkaloids in PB. Oripavine in PO varies from 0.02–0.15% and from 0.08–0.09% in capsules and roots, respectively. Oripavine occurs in PPO also but in micro amounts, as does isothebaine in PO. Levy et al. (1981) reported high heritability (h^2) of theabine content (42–44%) for 2 years based on parent-progeny regression (b_{op}) in PB; but h^2 for thebaine yield (mg/plant) was the lowest (0–14%), hence selection response (R) was also low.

2.2 Importance of Chemical Analysis

Modern alkaloid chemistry began with the isolation of the main components of opium obtained from *P. somniferum* (Szántay et al. 1983). The first semi-pure mixture of alkaloids was extracted with water from opium in 1803 by Derosne, who obtained a basic crystalline substance on precipitating the extract with potassium carbonate and called it "salt of opium"; Pelletier and Robiquet thought it to be "narcotine" (cf. Thakur 1983). Two years later, in 1805, Serturner isolated from opium the first vegetable-based crystalline "morphine", and assigned its basic chemical character. Subsequently, the introduction of preparative chromatography and associated spectroscopic instrumentation, such as sophisticated GC-MS, IR spectroscopy, etc. has opened the Pandora's Box of myriads of alkaloids occurring naturally not only in *P. somniferum* but also in various other taxa. As of now, more than 40 alkaloids have been isolated from opium (Šantavý 1970, 1979; Thakur 1983); apart from the six major alkaloids (morphine, codeine, thebaine, narcotine, papaverine, and narcine), all others occur in traces. The state of chemical analysis has reached such perfection that it made all this possible with considerable authenticity.

A rapid, cost-effective, and accurate chemical assaying of alkaloids in a reasonably good number of samples could also lead to the successful elucidation of the genetics/inheritance of opium alkaloids in *Papaver*, as suggested in the preceding section. Failure in precise chemical determination of these alkaloids could axiomatically jeopardize the factual genetic analysis. Hence, proper isolation and precise quantitation of alkaloids is an essential prelude to their meaningful genetic assaying. So important is the chemical analysis of alkaloids and for that matter, other substances.

3 Chemical Analysis of Major Opium Alkaloids

Consequent upon the need for genetic assaying of opium alkaloids for germplasm enhancement and for their efficient exploitation in pharmaceuticals, chemical analysis of these alkaloids is imperative, and entails chiefly the isolation and quantification of different alkaloids. The quantitation is further marked by separation (from each other) and measurement of each alkaloid. Further, methods of chemical analysis may be delineated into those specific for plant materials (crops), or classes of chemical compounds (alkaloids, steroids, glycosides, etc.). They may also be categorized in relation to technological and instrumental inputs. However, in this chapter, we present models which are, by and large, not specific to plant material, class of substance, or instrumentation tools.

3.1 Isolation of Alkaloids from Opium

Using the isolation procedure mentioned by Cromwell (1955), the major opium alkaloids and other bases are converted into their hydrochlorides by making a paste of opium with calcium chloride and extracting with warm water. Narcotine, papaverine, and thebaine are precipitated on treating the solution with 10% NaOH and morphine, codeine, and narceine remain in the alkaline solution. Codeine is then extracted by shaking with chloroform, and morphine is precipitated by making the remaining alkaline solution acidic and rendering it faintly alkaline with ammonia. Later, narceine is recovered from the remaining solution by evaporating the liquid to dryness and treating the residue with ethanol, while thebaine is separated as crystallized acid tartrate from the bases precipitated by sodium hydroxide. Narcotine and papaverine are separated from a solution in boiling water containing 0.33% oxalic acid, where the papaverine oxalate crystallizes on standing. Narcotine is finally precipitated with ammonia and crystallized by boiling ethanol.

3.2 Quantitation of Alkaloids

A variety of methods for quantitative determination of major opium alkaloids have been reported by different workers, such as gravimetric/volumetric methods, colorimetric methods, spectrofluorimetric methods, etc. However, separation is usually achieved by thin layer (TLC) or paper chromatography (PC). Today, advanced analytical methods, like gas (GLC)/high performance liquid chromatography (HPLC) are more frequently employed for this purpose. In the following sections, we describe these methods briefly, with greater emphasis on GLC and HPLC.

3.2.1 Gravimetric/Volumetric Methods

Among many gravimetric methods, the British Pharmacopoeia (1980) method of morphine analysis (in opium) is carried out in the following four sequential steps: (1) making a slurry of opium in water with $Ca(OH)_2$ and thus converting the resinous matters into insoluble lime compounds; decomposition of the morphine meconate with formation of insoluble calcium meconate, and dissolution of the resulting free morphine in lime; (2) decomposition of the solution by ammonium chloride with formation of calcium chloride, ammonia, and free morphine; (3) dissolving the impurities by using alcohol and promoting the crystallization of alkaloid by ether; and (4) estimation of morphine titrimetrically.

3.2.2 Colorimetric Methods

Many colorimetric methods have been reported to estimate opium alkaloids; for instance, the method of Adamson based on the Radulescu color reaction using a violet filter; the method of Gramer based on the reaction with nickel sulfate using a red filter for color determination; Moorhoff's method using ferric chloride; and Wegner's method involving a reaction with diazobenzene sulfonic acid; etc. (Cromwell 1955). However, after 1960, more efficient methods were standardized and employed for alkaloid estimation in opium (Hiroshi 1961; Hiroshi et al. 1962; Prista et al. 1976; Chichiro and Suranova 1978; Vaidya et al. 1980). Of these, the method of Vaidya et al. (1980) is a rapid method for extraction and colorimetric estimation of morphine in opium. It involves the treatment of a methanol extract of opium with activated charcoal to obtain a colorless extract of all the major opium alkaloids. Morphine content is estimated colorimetrically by using H_2O_2-Cu-NH_3. The reaction is specific to morphine, as no other major opium alkaloid responds to this method.

3.2.3 Spectrofluorimetric Determination

Spectrofluorimetric method was reported by Chalmers and Wadds (1970) for the estimation of opium alkaloids in micro amounts where morphine, codeine, papaverine and narcotine mixtures are determined spectrofluorimetrically in 2.5 h with $\leq 7\%$ relative errors. Morphine and codeine are determined by a differential method involving fluorescence measurements in 0.1 N sulfuric acid and 0.1 N sodium hydroxide solutions, after the papaverine and narcotine were extracted into chloroform from the sample solution in 0.1 N sulfuric acid. Then papaverine and narcotine are determined after extraction at pH 9. Trichloroacetic acid is used for simultaneous enhancement of the narcotine fluorescence and quenching of the papaverine fluorescence. The standard used is 2-aminopyridine.

Later, Dumitrashko and Babilev (1976) separated morphine and codeine by the combined methods of PC and fluorimetry. First, they separated them through paper chromatography (see Sect. 3.2.4) with iso-BuOH-HOAC-H_2O (100:10:25) as developing solvent, followed by their quantitative determination fluorimetrically using 285 nm as the excitation wavelength and 345 nm as the emission wavelength. The standard curve remains linear in the range of 5–20 μg/ml for each drug. However, fluorimetric methods are not frequently used.

3.2.4 Thin Layer (TLC) and Paper Chromatographs (PC)

Opium alkaloids are most commonly estimated by separating them on a chromatographic paper (PC) and/or on a thin layer chromatographic plate (TLC). The former needs a longer time for developing spots than on the silica gel layers of TLC. After the separation of different opium alkaloids with different mobile phase systems of PC and TLC, individual alkaloids are quantitatively measured by eluting the spot and using different procedures, like spot area measurement, spectrophotometry, IR spectroscopy, densitometry, etc. Important methods are described below:

3.2.4.1 Thin Layer Chromatography (TLC)

Mary and Brochmann-Hanssen (1963) developed a sound method of TLC coupled with spectrophotometry for quantitative analysis of opium alkaloids in a single extraction. The method runs as follows.

Preparation of Opium Extracts. 200 mg of finally powdered opium is triturated to a paste in a glass mortar and pestled with 0.5 ml methanol and 0.5 ml concentrated ammonium hydroxide solution. Then 3 g of aluminum oxide is added and the trituration continued until a homogeneous free-flowing powder is obtained. The triturate is then transferred quantitatively to a chromatographic column (30 × 1 cm). The alkaloids are eluted with 80 ml of a mixture of chloroform-isopropyl alcohol (3:1). The rate of flow is adjusted to about one drop per second. The eluate is finally concentrated almost to dryness under vacuum and then transferred to a small vial to evaporate to dryness in a stream of nitrogen to obtain the opium extract.

Standard Solution and TLC. A 0.25% (w/v) of each of the alkaloid bases in MeOH:CHCl₃ (1:4) is taken as standard alkaloid solution. MeOH:CHCl₃ (1:9) is used as developing solvent for morphine, codeine, and thebaine, and EtOH:C₆H₆ (1:4) for papaverine and narcotine TLC on silica gel

G. Varying amounts (0.01–0.05 ml) of standard alkaloid solution of each alkaloid are spotted on a TLC plate. After developing the plate, each alkaloid is scraped off and dissolved in MeOH to make up to 5 ml. The absorbance is determined for each alkaloid by spectrophotometer against a MeOH blank at the following wavelengths: morphine – 286 nm, codeine – 215 nm, thebaine – 285 nm, papaverine – 279 nm, and narcotine – 312 nm. Potassium iodoplatinate reagent/UV light is used to visualize the spots. A standard curve is prepared for each alkaloid by plotting the amount of each alkaloid against its absorbance.

Sample Solution and TLC. The alkaloid extract is dissolved in exactly 5 ml of MeOH:CHCl$_3$ (1:4), and five 0.05 ml volumes of this solution are spotted on each of two plates coated with silica gel G. One plate is developed with MeOH:CHCl$_3$ (1:9) for determination of morphine, codeine, and thebaine; while the other plate is developed with EtOH-C$_6$H$_6$ (1:4) for papaverine and narcotine determination. The silica gels of various alkaloids are scrapped and made up to 5 ml with MeOH and the absorbance is measured by spectrophotometer as above. The percentage of each alkaloid is finally calculated by using the standard curve of each alkaloid.

3.2.4.2 Paper Chromatography

Many methods of PC in combination with other methods were developed for opium alkaloid analysis. Three of them are briefly described here:

1. Five major opium alkaloids: morphine, codeine, thebaine, papaverine and narcotine, were determined by the densitometric method (Haruyo and Masako 1957) after paper chromatography based on the Rf, wavelength of the maximum, and the range of linearity between the amount of alkaloid and absorption coefficient (given, in order, in parentheses) as follows: morphine (0.72, 286 nm, 10–60γ) and codeine (0.83–0.5, 283 nm, 5–30γ) were chromatographed by using the upper layer of the BuOH – 28% NH$_3$-H$_2$O (50:9:15); while thebaine-HCl (0.50–0.4, 285 nm, 10–60γ) and papaverine-HCl (0.68–0.70, 281 nm, 5–40γ) were developed by the water-saturated butanolic 4 vol% acetic acid. Narcotine (0.69, 290 nm, 10–60γ) was chromatographed by the upper layer of BuOH-AcOH-H$_2$O (5:1:4).

2. Morphine, codeine, papaverine, and narcotine were separated by Balatre et al. (1960) by paper chromatography on paper buffered to pH 3.5 with a solution of 3.5% K$_2$HPO$_4$ and 2% citric acid (3:7) using a developing solvent involving toluene: iso-BuOH saturated with water (1:1). The cut-out spot of each alkaloid was leached with a small volume of acetic acid for 24 h. The AcOH solution was heated with concentrated sulfuric acid, excess ammonia was added, and the fluorescence measured. Then

the percentage of alkaloids was determined with the help of standard curves.

3. Thebaine, papaverine, and narcotine were separated by chromatography (Izmailov et al. 1964) on paper treated with buffer solution (prepared from 0.1 M citric acid and 0.2 M NaH_2PO_4) using diethyl ether (saturated with water for 1.5–2 h) as developing solvent. The spots containing the alkaloids were eluted with methanol and the absorbance of the eluates were determined at 222, 223, and 240 nm for narcotine, thebaine, and papaverine, respectively. The amounts of alkaloids in the eluates were calculated from the calibration curve. For morphine and codeine estimation, earlier Izmailov et al. (1963) adopted the following method: 10 ml of water was added to 5 ml poppy extract. The mixture was evaporated to 7.5 ml, acidified with diluted H_2SO_4, again evaporated to 2–3 ml, filtered (pH adjusted to 10 by using 1% NaOH), and then extracted with diethyl ether. This ether extract was concentrated to 0.5–1 ml for codeine estimation; for morphine determination, the alkaline aqueous fraction was acidified with diluted H_2SO_4 (pH adjusted to 9–10 with 10% NH_4OH) extracted with EtOH-CHCl$_3$ (1:2) and then evaporated to 0.5–1 ml. Absorbance was measured at 220 nm. Finally, the morphine and codeine were estimated with the help of standard curves.

3.2.5 Gas Liquid Chromatography (GLC)

Gas chromatography, used by many workers for opium alkaloids' estimation, is particularly valuable because of its speed, simplicity, and reproducibility. The following procedures were employed under GLC by different workers to separate and quantify opium alkaloids:

1. A reliable GLC method was developed by Fisher and Gillard (1977) for analyzing papaveretum – a synthetic mixture of the hydrochlorides of the opium alkaloids, containing the equivalent of anhydrous morphine (47.5–52.5%), anhydrous codeine (2.5–5.0%), noscapine (narcotine, 16–22%), and papaverine (2.5–7.0%). Since this method can well be extended to opium also, it is described in detail. It involves derivatization of the alkaloid hydrochlorides in papaveretum without prior extraction, followed by direct injection into the gas chromatograph. Standards of morphine, codeine, papaverine, and noscapine are dried at 120 °C for 2 h immediately before use. The analysis is performed on a gas chromatograph equipped with dual-flame ionization detectors and a 5-nm recorder with a chart speed of 10 mm/min. The columns used are 120 cm × 3 mm (o.d.) glass-lined metal packed with 2% OV-101 on 100–120 mesh Gas Chrom Q. The injector and detector temperature is maintained at 300 °C, while the oven temperature is maintained initially at

180 °C for 5 min, then programmed from 180–250 °C at the rate of 7°/min, and then held at 250 °C for 3 min. The nitrogen gas is used as carrier gas at a flow rate of 30 ml/min; the air and hydrogen flow rates are adjusted to give optimum detector response. The packed columns are conditioned overnight at 325 °C.

Following are the procedures for preparation of standard and sample solutions and their chromatographic analysis:

Standard Solution. Place 200 mg of morphine and 80 mg of noscapine into a 50-ml volumetric flask to prepare solution A. Similarly, 80 mg each of codeine and papaverine are placed in a 200-ml volumetric flask to develop solution B. The volumes in both solutions are made up with anhydrous pyridine. Now pipet 1 ml of solution A and solution B each into a 10-ml conical flask. Add 1 ml acetic anhydride and allow to stand for 20 min. Evaporate to dryness with a stream of dry introgen and add 0.5 ml internal standard solution prepared by dissolving 100 mg squalene in 100 ml of 95% ethyl acetate and 5% acetic acid. Inject 0.5 µl of this solution into the gas chromatograph. Eventually, the area of each compound is recorded using an electronic digital integrator.

Sample Solution. This is prepared by accurately weighing 170 mg of the papaveretum sample into a 50-ml volumetric flask and diluting with anhydrous pyridine. Now pipet 2 ml aliquot of this solution into a 10-ml flask and then proceed as above in the case of standard solution.

Estimation of the Compound. The amount of the respective alkaloid in the sample is calculated from the following equation:

$$Ca = \frac{Rs \times Cst}{Rst \times Cs \times 2} \times 100,$$

where,

Ca = percent (w/w) of the particular alkaloid base
Cst = concentration of alkaloid in the standard solution in mg/ml
Cs = concentration of papaveretum in the sample solution in mg/ml

$$Rs = \frac{\text{peak area of particular component in sample chromatogram}}{\text{peak area of internal standard in sample chromatogram}}$$

$$Rst = \frac{\text{peak area of particular component in standard chromatogram}}{\text{peak area of internal standard in standard chromatogram}}.$$

Other methods used with some modified conditions are as follows:

2. Brochmann-Hanssen and Furuya (1964) separated opium alkaloids on SE-30 column under the following chromatographic conditions: carrier

gas–argon; inlet pressure, p.s.i., lb-19; U-shaped glass column 180 cm × 3.13 mm (6′ × 1/8″) column temp. – 207 °C; injection part temp. – 302 °C; cell-bath temp. – 230 °C. The alkaloid samples were dissolved in acetone and applied with a microsyringe.

3. Nieminen Elna (1971) devised a method which entails separation of samples by liquid-liquid extraction with (3:1) $CHCl_3$-iso-PrOH, followed by quantitative determination of morphine, codeine, papaverine, narcotine, and thebaine on a SE-30 column using histapyrodine-HCl and estradiol valerate as internal standard.

4. A faster gas-chromatographic method was developed by Furmanec Dmytro (1974) for the simultaneous determination of morphine, codeine, thebaine, papaverine, and narcotine alkaloids based on complete separation on a column packed with a 50:50 mixture of OV-17 phenyl-methylsilicone and SE-30 silicone gum on chromosorb W. Standard deviation for the individual alkaloid was between 0.05 to 0.18%.

5. In yet another method, Ono et al. (1977, 1978) separated opium alkaloids on a silicone OV-1 column as follows: after triturating with 1% HCl, opium was filtered and extracted with $CHCl_3$. The $CHCl_3$ extract was treated with NH_3, evaporated and dissolved in acetone, and injected in gas chromatograph for thebaine, papaverine, and noscapine determination. The aqueous acidic layer was neutralized with NH_3 and extracted with $CHCl_3$-iso-PrOH (3:1) mixture, evaporated, and the residue was treated with bistrimethylsilylacetamide at 90 °C for 10 min. Codeine and morphine were then determined by gas chromatography.

As is apparent from the foregoing, most of the procedures frequently lack the sensitivity and specificity required for routine analysis of opium alkaloids: the official methods (gravimetric/volumetric) are tedious and time-consuming; the PC and TLC procedures are also tedious and unsatisfactory because of overlapping spots. Further, GLC, although frequently used, is fraught with the problem of morphine adsorption on the GC column, and the decomposition of thebaine at high temperature. Therefore, a sensitive and faster method was needed to assay major opium alkaloids. The high performance liquid chromatography (HPLC) method holds great promise for such a rapid and accurate analysis, especially when combined with a simplified method of sample preparation. In the following section, we describe HPLC in detail.

3.2.6 High Performance Liquid Chromatography

Although HPLC methods are available for individual opium alkaloids, it is always convenient to separate and estimate all the major opium alkaloids in a single injection experiment. Both normal and reverse phase columns are applied under HPLC analysis. However, the methods based on normal phase liquid chromatography (Beasley et al. 1974; Vincent and Engelke 1979; Hodges and Rapoport 1982) are unsuitable for routine analysis of opium because sample preparation is time-consuming, extraction is not complete, and the method is highly sensitive to water content in the sample, mobile phase, and column adsorbant (Srivastava and Maheshwari 1985). Therefore, methods based on reversed phase liquid chromatography are attempted most frequently for alkaloid analysis in both latex and poppy straw as elaborated below:

3.2.6.1 HPLC for Alkaloid in Gum Opium

The following two standardized methods of HPLC have been found to be useful for gum analysis of opium alkaloids:

1. *Method of Nobuhara et al.* (1980). A simple and rapid method for the routine quantitative analysis of the six major alkaloids in gum opium by direct isocratic HPLC on a reversed phase partition mode column without using ion-pair reagents was successfully used by Nobuhara et al. (1980) under the following conditions:

Apparatus. (1) A Waters' Model ALC/GPC 204 liquid chromatograph, equipped with a Model 6000 A pump, (2) a Model 440 detector (254 nm), (3) a Model U 6 K injector, and (4) columns – stainless-steel tubes (300 × 4 mm i.d.), packed with Nucleosil 10 CN or Nucleosil 10 C_{18}.

Mobile Phase. A mixture of 1% ammonium acetate buffer (required pH adjusted with acetic acid) and acetonitrile; or the buffer with acetonitrile and dioxane is used as mobile phase.

Alkaloid Standard Solution. Dissolve 10 mg each of morphine, codeine, cryptopine, thebaine, narcotine, and 20 mg of papaverine in MeOH, and make up to 20 ml.

Determination of Alkaloids in Gum Opium. Shake 2 g gum opium mechanically with 2.5% acetic acid (20 ml × 3) for 20 min every time. Then centrifuge the extract, separate, and filter the supernatent. Combine the extracts and make up to 100 ml with 2.5% acetic acid. Dilute 5 ml volume of this to 20 ml with MeOH.

Now inject 6 µl of the solution into the liquid chromatograph.

The Nucleosil 10 CN column is more suitable than the Nucleosil 10 C_{18} column because the separation is rapid and complete and the morphine is eluted far enough from the solvent front. Retention times of opium alkaloids on CN column by using mobile phase – 1% ammonium acetate (pH 5.8)-acetonitrile-dioxane (80:10:10) at a flow rate of 1.5 ml/min is: morphine

4.1, codeine 5.1, cryptopine 8.1, thebaine 9.2, narcotine 12.3, and papaverine 15.7 min.

The alkaloid content is now calculated by the calibration curve for each alkaloid and measurement of the height of the corresponding peak in the chromatogram of the extract. The coefficient of variation of the method by repeated analysis for the individual alkaloid is reported to be <1.5%.

2. *Improved/Modified HPLC Method.* Srivastava and Maheshwari (1985) compared the available liquid chromatographic methods for opium gum analysis and developed a modified method which is more sensitive and faster than others. They used five extractions with 2.5% aqueous acetic acid instead of four reported earlier (Nobuhara et al. 1980) for the complete extraction of major opium alkaloids. They also modified the mobile phase of Nobuhara et al. (1980). The modified one produced better resolution of peaks for the five alkaloids and the chromatographic separation was complete within 15 min. Recovery was 100% for codeine, thebaine, and narcotine, and 98% for morphine and papaverine under the following operating conditions:

Apparatus.
1. Waters' Associates Model ALC/GPC-244, equipped with M-6000A pump
2. U6K injector
3. M-440 absorbance detector
4. Omniscribe B-5000 recorder
5. Shimadzu data processor
6. Reversed phase μ-Bondapak-CN column (30 cm × 3.9 mm i.d. packed with 10-μm particles).

LC Operating Conditions.
1. Mobile phase comprising 1% sodium acetate in water (pH 6.78, adjusted with glacial acetic acid)-acetonitrile-1,4-dioxane (75:20:5) at the flow rate of 1.5 ml/min
2. UV detector at 254 nm; 0.1 aufs (absorbance units full scale); chart speed 0.5 cm/min at temperature 25 °C.

Standard Solution. Dissolve 10 mg of each of the standard morphine, codeine, thebaine, and narcotine and only 5 mg of papaverine in 25 ml MeOH to obtain the standard solutions. Inject varying amounts: 0.4, 0.8, 1.2, 1.6, 2.0, 2.4, 3.2, 4.0, and 4.8 µg of all the standard solutions (up to only 2.0 µg in the case of papaverine) into the LC system to obtain the peaks of standards.

Sample Preparation. Shake mechanically the finally powdered and dried 1 g gum opium (dried over anhydrous $CaCl_2$ in desiccator at $\simeq$40 °C) for 20 min each time with 2.5% aqueous acetic acid (20 ml × 5). Decant the mixture, filter into 100-ml volumetric flask and make the final volume up to 100 ml with 2.5% acetic acid. Dilute 10 ml of above extract to 25 ml with MeOH, and filter through a 0.45 µm Millipore membrane filter.

Inject 2–5 µl of aliquot into the LC system.

Calibration Curves. Standard curves are drawn by plotting peak heights against the quantity of the alkaloid injected. Calibration curves are linear up to 4 µg for morphine, 3.2 µg for codeine, 4 µg for thebaine, 4 µg for narcotine, and 1.6 µg for papaverine.

The percentage of each alkaloid in the sample is calculated by comparing peak heights with those for respective standards.

3. *Miscellaneous Methods.* Many more HPLC models are reported of alkaloid estimation prior to and after the above two methods (Ziegler et al. 1975; Lurie 1977; Wu and Wittick 1977; Matantseva et al. 1980; Hutin et al. 1983; and so forth). In particular, however, the method developed by Staba et al. (1982) for determination for alkaloids in tissue culture *Papaver* samples (using MeOH: 0.3% ammonium carbonate in water in 75:25 ratio, as eluting solvent at 254 nm) is of interest. It improves upon the recovery time of Nobuhara et al.s' (1980) model for thebaine (8.2 min vs. 9.2 min) and papaverine (5.5 min vs. 15.7 min), but delays recovery of the remaining alkaloids.

Other methods employing reversed phase µ Bondapak C_{18} column with UV detector at 254 nm or twin mobile phases are generally not so efficient for quantitation of *Papaver* alkaloids.

3.2.6.2 HPLC for Alkaloids in Poppy Straw

In view of social abuses of gum opium on a global scale, emphasis has now been shifted to the cultivation of *Oxytona* species, particularly *Papaver bracteatum*, containing only non-narcotic thebaine in straw, as a viable substitute to the opium poppy (*P. somniferum*). Even in countries like India where opium poppy is grown for opium, notwithstanding the strict governmental control, harvesting of poppy straw is gaining ground. As the straw also contains opium alkaloids, although in very small quantity, a more sensitive and accurate method than that for gum opium is needed for alkaloid assaying: the gum methods are unable to extract the alkaloids completely from the straw. Some of the methods which were developed by different workers are elucidated here:

1. *Phenyl Bonded Reversed Phase HPLC.* Employing a simple and rapid system of sample preparation, Pettitt and Damon (1982) developed a reversed phase HPLC for straw analysis. This method operates under the following conditions of LC:

Apparatus.
1.   A Waters' Associates model 201, equipped with 6000 A pumps
2.   A Model 660 solvent programmer
3.   A U 6 K injector
4.   A Perkin Elmer (Coleman) Model LC-55 variable wavelength detector.

The Perkin-Elmer Model I computing integrator provides areas and retention times. The column is a phenyl Bondapak (Waters'), 25 cm × 5 mm (i.d.) used in conjunction with a 7 cm × 2 mm (i.d.) guard column packed with C_{18}/Corasil (Waters'), particle size 37–50 μm. The column eluted is monitored at 275 nm.

Mobile Phase. Acetonitrile-water (20:80) containing 1 ml/l of glacial acetic acid and 0.04 ml/l of N,N-dimethyloctylamine (pH adjusted to 3.5 with NaOH) with a flow rate of 1 ml/min.

Solvent A. Acetonitrile-water (5:95) containing 1 ml/l glacial acetic acid and 0.04 ml/l N,N-dimethylamine (pH 3.5, adjusted with NaOH), with quinine sulfate as the internal standard.

Sample Preparation. Weigh accurately 50 mg of powdered sample of straw in a small Erlenmeyer flask. To this, add 25 ml of the solvent A. The solution is then effected by sonication for 30 min. Filter the aliquot of this solution through a Millipore filter (pore size 0.45 μm).
   Standards are dissolved in the same batch of solvent plus internal standard as for the sample solution.
   The phenyl column gives superior peak shapes compared to either the octyldecyl or octyl type because the presence of N,N-dimethyloctylamine eliminates tailing in peaks.

2. *Bondapak C_{18} Column Reversed Phase HPLC.* A method of quantitative determination of opium alkaloids on Bondapak C_{18} column was standardized by Akhila and Uniyal (1983) at our Institute. It operates under the conditions noted below:

Apparatus Required.
1.   A Waters' modular liquid chromatograph equipped with a Model 6000 A pump
2.   A Model 450 variable wavelength detector (at 254 nm)
3.   A Model U6K injector
4.   Columns of stainless steel tubes (300 × 4 mm i.d.), packed with μ-Bondapak C_{18}.

Mobile Phase. Sodium hydrogen phosphate (3.9 g) is dissolved in redistilled water (190 ml) and made up to 250 ml with acetonitrile to give a solution of

0.1 N strength of pH 4.8. This solution is then passed through Millipore filters with a pore size of 0.45 μm.

Alkaloid Standard Solution and Calibration Curve. Morphine (10 mg), thebaine (2 mg), papaverine (2 mg), codeine (2 mg), crytopine (2 mg), and narcotine (6 mg) are dissolved in methanol and made up to 5 ml in a volumetric flask. Calibration curve for each alkaloid is prepared by injecting varying amounts (microliters) of standard solutions into the liquid chromatograph, and measuring the peak height against the amount injected.

Preparation of Concentrated Poppy Straw (CPS) and Alkaloid Determination. Dried poppy straw (20 g) is extracted in a Soxhlet apparatus with MeOH (12 h), concentrated to a pasty mass and treated with 0.25 g $Ca(OH)_2$ and 2.5% acetic acid solution (5 ml). The mixture is centrifuged and the supernatant separated and filtered. This process of extraction is repeated three times to obtain the CPS, which is then made up to 10 ml solution in a volumetric flask with the acetic acid solution.

One μl of this solution is injected for separation of alkaloids on liquid chromatograph, where morphine (5.03 min), codeine (6.03 min), cryptopine (7.73 min), thebaine (11.86 min), papaverine (20.35 min), and narcotine (23.28 min) are eluted successively.

Alkaloid percentage is finally calculated by relative height of peak against that of the corresponding standard.

3. *CN Column Reversed Phase HPLC.* Another improved method of straw analysis was recently developed by Verma et al. (1990) at our Institute. This method gives better recovery of alkaloids when the sample is extracted with 2.5% aqueous acetic acid in combination with lime, instead of aqueous acetic acid alone and when the mobile phase of 1% sodium acetate in water (pH adjusted to 7.4 with glacial acetic acid)-acetonitrile-tetrahydrofuran (80:16:4) is used rather than the mobile phase of 1% sodium acetate in water (pH adjusted to 6.78 with glacial acetic acid)-acetonitrile-1,4-dioxane (75:20:5), used by Srivastava and Maheshwari (1985).

Apparatus.
1. Waters' Associates Model, equipped with two 6000 A pumps
2. A U6K injector
3. 450 variable wavelength detector
4. A data processor, Waters' 730 to process areas and retention times
5. The column reverse phase μ Bondapak CN (Waters') column, 30 cm × 3.9 mm (i.d.).

LC Operating Conditons. Mobile phase flow rate 0.7 ml/min: UV detector at 254 nm, 0.04 aufs, Chart speed 0.6 cm/min, temperature 25 ± 1 °C.

Sample Preparation. Air-dried poppy straw (10 g) is extracted with MeOH (5 × 100 ml), concentrated in vacuo and treated with calcium hydroxide (0.1 g) and 2.5% acetic acid (5 ml) in order to make a slurry. The slurry is stirred for 30 min, homogenized, and centrifuged. The supernatant is separated, filtered, and the volume of sample solution is made up to 10 ml with acetic acid.

HPLC Analysis. One µl of the aliquot, filtered through a Millipore filter (pore size 0.45 µm) is injected into the LC system. Morphine, codeine, thebaine, papaverine, and narcotine are eluted at the retention times of 12.16, 16.20, 21.09, 25.64, and 29.16 min, respectively. Alkaloid percentage is then estimated with the help of linear calibration curves of these alkaloids, obtained by plotting peak area against the quantity of alkaloid injected. Recovery of these alkaloids ranges from 95–100%.

4. *HPLC for* Oxytona *Alkaloids.* Milo et al. (1988) applied a reversed phase HPLC method for simultaneous quantitation of the alkaloids: salutaridine, thebaine, oripavine, alpinigenine, isothebaine, and orientalidine of section *Oxytona* of the genus *Papaver* as follows:

Extraction of the Alkaloids. Accurately weigh 250 mg of dried powdered plant material, and extract it with 5% aqueous acetic acid (20 ml, 18 h at room temp.). Filter the solution and add hexane (7 ml). The aqueous fraction is then made alkaline with ammonium hydroxide, and extracted with a 3:1 mixture of chloroform-2-Propanol (3 × 15 ml); the pooled extracts evaporated to dryness under the reduced pressure; the residue dissolved in 2.5 ml of MeOH and filtered through a RC 55, 0.45-µm membrane filter.

HPLC Analysis. A Tracor liquid chromatograph equipped with a Model 951 pump unit, a 980 A solvent programmer, a 970 A detector and a computing integrator Model CI-10B (LDC-Milton Roy) are used for analysis. Alkaloids are finally separated by reversed phase HPLC carried out on a LiChrosorb Superspher RP-18 column (Merck, particle size 4 µm, 125 mm × 4 mm i.d.) at 25 °C by the mobile phase 2-propanol-acetonitrile-water (5:40:50) with 1% ammonium carbonate (flow rate 1 ml/min, UV detector at 280 nm).

5. *Miscellaneous Methods.* Two other methods may also be mentioned for their usefulness in shortening the elution time and novelty in the extraction process. Using reversed phase HPLC with UV detector and twin mobile phases A and B, Dogan et al. (1986) developed a method of alkaloid estimation. The mobile phase B, i.e., 0.05 M sodium acetate buffer (pH 3.6 adjusted with acetic acid)-acetonitrile-THF-EtOH (88:6:3:3) gave quicker results than the mobile phase A, i.e., 0.05 M sodium acetate buffer (pH

3.6)-acetonitrile-EtOH (88:8:4) for the separation of major opium alkaloids. The retention time is dependent upon the pH of the buffer and the mobile phase. Addition of THF into the mobile phase shortens the elution time considerably without affecting the resolution of alkaloids. The HPLC instrument used by them was a Waters' Associates Model 204 LC equipped with Model 6000 A solvent delivery pump, Model 440 UV/VIS detector (254–658 nm) and a Model U6K injector. Data recording and processing was carried out with Perkin-Elmer Model 56 Recorder and a Pye Unicam CDP1 computing integrator. The column was prepacked with 10 μm particles μ-Bondapak C_{18} (Waters' Assoc.) and caffeine was used as the internal standard.

Recently, another method of alkaloid extraction followed by HPLC analysis was suggested by Srivastava et al. (1989). They suggested extraction of dried powdered capsule husk (1 g) with 5% glacial acetic acid (3 × 25 ml) for 20 min, always using a shaker. The extract is then filtered and residue reextracted three times as before. The total extract (pH adjusted to 9.0–9.5 with 25% aqueous ammonia solution) is now extracted with a mixture (4 × 15 ml) of chloroform and isopropanol (3:1), dried over anhydrous sodium sulfate, evaporated to dryness under vacuum at 50 °C, redissolved in MeOH (5 ml) and analyzed over HPLC under the same LC conditions as described by Srivastava and Maheshwari (1985) for gum analysis (see method 2).

4 Concluding Remarks

The Egyptian *Spnn-Spnn*, The Arabian *Abou-el-noum*, the Chinese *minans*, the Indian *afim*, and the primitive *Megisterium opii* are all the same, i.e., todays' *opium* – a word coined by the ancient Greeks. Opium reached a peak of scientific significance because the isolation of the first alkaloid from opium ex-*P. somniferum* marked the beginning of modern alkaloid chemistry. (Szántay et al. 1983). This initiated the genetic and chemical investigations.

With the growing demand for rapid chemical analysis to quantitate alkaloids, an array of chemical methods, each an improvement over the past, has been developed for five major opium alkaloids occurring in measurable quantity in *Papaver*. Among these, the reversed phase HPLC with a single run and with minimal sample preparation time is an easy and reproducible method for routine analysis of *Papaver* alkaloids.

However, chemical analysis of minor opium alkaloids, exposed to a number of commonly occurring enzymes in latex and synthesized in microquantities, is extremely difficult. Hence, genetic analysis of such minor alkaloids has hitherto been largely avoided.

References

Akhila A, Uniyal GC (1983) Quantitative estimation of opium alkaloids. Indian J Pharm Sci 45:236–238

Asahina H, Kawatani T, Ono M, Fujita S (1957) Studies on poppies and opium. Bull Narc 9:20–33

Balatre P, Traisnel M, Delcambre JP (1960) Paper chromatography of opium alkaloids. Bull Soc Pharm Lille: 143–149 (CA 1962, 56:10288a)

Beasley TH, Smith DW, Ziegler HW, Charles RL (1974) Liquid-chromatographic system for separating opium alkaloids. J Assoc Off Anal Chem 57:85–90

Böhm H (1967) Über *Papaver bracteatum* III. Mitteilung. Charakteristische Veränderung des Alkaloidspektrums während der Pflanzenentwicklung. Planta Med 15:215–220

British Pharmacopoeia London (1980) Her Majesty's Stationary Office England, vol I, p 316

Brochmann-Hanssen E, Furuya T (1964) Opium alkaloids – separation and identification by gas, thin layer and paper chromatography. J Pharm Sci 53:1549–1550

Chalmers RA, Wadds GA (1970) Spectrofluorimetric analysis of mixtures of the principal opium alkaloids. Analyst 95:234–241

Chelombit'Ko VA, Mikheev AD (1988) Data on chemotaxonomy of the section Macrantha Elkan of the genus *Papaver* L. Part I. Alkaloids of *Papaver bracteatum* Lindl. Rastit Resur 24:400–410 (PBA 59:1407, 1989)

Chichiro VE, Suranova AV (1978) Unified method for the determination of morphine in opium preparations. Farmatsiya (Mosc) 27:76–78

Cromwell BT (1955) The alkaloids. In: Paech K, Tracey MV (eds) Modern methods of plant analysis, vol IV, Springer, Berlin Heidelberg Göttingen, pp 448–460

Dogan M, Gevenkiris A, Tarhan O (1986) The analysis of poppy straw concentrate by high-performance liquid chromatography. Fitoterapia 57:257–261

Dumitrashko AI, Babilev FV (1976) Spectrofluorimetric method for determining morphine and codeine in opium poppy pods. Izv Akad Nauk Mold SSR, Ser Biol Khim Nauk 1:83–84

Fairbairn JW (1976) Conference on medicinal plants, Marianske Lazne (CSSR), 1975. Planta Med 29:26

Fisher G, Gillard R (1977) GLC determination of opium alkaloids in Papaveretum. J Pharm Sci 66:421–423

Furmanec Dmytro (1974) Quantitative gas chromatographic determination of the major alkaloids in gum opium. J Chromatogr 89:76–79

Goldblatt P (1974) Biosynthetic studies in *Papaver* section *Oxytona*. Ann Mo Bot Gard 61:264–296

Haruyo A, Masako O (1957) A unified analyses of opium for main alkaloids by paper chromatography. Eisei Shikenjo Hokoku 75:133–139

Hiroshi S (1961) Analysis of opium alkaloids. XI. A colour reaction of codeine and morphine and its application to colorimetric determination. Yakugaku Zasshi 81: 865–869

Hiroshi S, Kozuo S, Takeshi T (1962) Analysis of opium alkaloids and allied compounds. XV. Colorimetric determination of dihydrocodeine by ceric ammonium nitrate and 2,4-dinitrophenylhydrazine. Yakugaku Zasshi 82:1684–1687

Hodges CC, Rapoport H (1982) Morphinan alkaloids in callus cultures of *Papaver somniferum*. J Nat Prod 45:481–485

Husain A, Sharma JR (eds) (1983) The opium poppy: medicinal and aromatic plants series 1. Central Institute of Medicinal and Aromatic Plants (CIMAP), Lucknow (India)

Hutin M, Cave A, Faucher JP (1983) Utilization of higher-performance liquid chromatography in the characterization and determination of alkaloid principles of *P. somniferum* L. J Chromatogr 268:125–130

Izmailov NA, Mushinskaya SKh, Danel'yants VA (1963) Paper chromatography for controlling the manufacture of alkaloids from the capsules of oleiferous poppy. II. Determination of morphine and codeine. Med Prom SSSR 17:47–50

Izmailov NA, Mushinskaya SKh, Danel'yants VA (1964) Paper chromatography for controlling the manufacture of alkaloids from the capsules of oleiferous poppy. III. Determination of narcotine, thebaine and papaverine. Med Prom SSSR 18:47–51

Khanna KR, Shukla S (1986) HPLC investigation of the inheritance of major opium alkaloids. Planta Med 52:157–158

Laane MM, Wold JK, Paulsen BS, Haugli T, Nordal A (1988) Cytogenetic studies in "X-7", a new line of *Papaver bracteatum* Lindl. with increased thebaine content. Hereditas 108:187–197

Lal RK (1988) Genetics of economic traits related to primary and secondary metabolic activities in opium poppy (*Papaver somniferum* L.). PhD Thesis, Kanpur University, Kanpur

Lal RK, Sharma JR (1991) Genetics of alkaloids in *Papaver somniferum*. Planta Med 57:271–274

Lalezari I, Nasseri P, Asgharian R (1974) *Papaver bracteatum* population Arya II. J Pharm Sci 63:1331–1332

Levy A (1985) A shattering resistant mutant of *Papaver bracteatum* Lindl.: Characterization and inheritance. Euphytica 34:811–815

Levy A, Milo J (1991) Inheritance of morphological and chemical characters in interspecific hybrids between *Papaver bracteatum* and *P. pseudo-orientale*. Theor Appl Genet 81:537–540

Levy A, Palevitch D, Lavie D (1981) Genetic improvement of *Papaver bracteatum*: heritability and selection response of thebaine and seed yields. Planta Med 43:71–76

Lurie I (1977) Application of reverse phase ion-pair partition chromatography to drugs of forensic interest. J Assoc Off Anal Chem 60:1035–1040

Mary NY, Brochmann-Hanssen E (1963) Determination of the principal alkaloids of opium by thin layer chromatography. Lloydia 26:223–228

Matantseva EF, Gladyshev PP, Goryaev MI, Bektenova GA (1980) Quantitative determination of opium alkaloids by liquid chromatographic methods. Khim Prir Soedin 5:730–731

Milo J, Levy A, Palevitch D, Ladizinsky G (1987) Thebaine content and yield in induced tetraploid and triploid plants of *Papaver bracteatum*. Euphytica 36:361–367

Milo J, Levy A, Ladizinsky G, Palevitch D (1988) Phylogenetic and genetic studies in *Papaver* section *Oxytona*: cytogenetics, isozyme analysis and chloroplast DNA variation. Theor Appl Genet 75:795–802

Milo J, Levy A, Palevitch D, Ladizinsky G (1990) Genetic evidence for the conversion of the morphinan alkaloid thebaine to oripavine in interspecific hybrids between *Papaver bracteatum* and *P. orientale*. Heredity 64:367–370

Nieminen E (1971) Gas-chromatographic determination of opium alkaloids in pharmaceutical products. Farm Aikak 80:342–349

Nobuhara Y, Hiráno S, Namba K, Hashimoto M (1980) Separation and determination of opium alkaloids by high-performance liquid chromatography. J Chromatogr 190: 251–255

Nyman U, Bruhn JG (1979) *Papaver bracteatum* – summary of current knowledge. Planta Med 35:97–117

Nyman U, Hall O (1976) Some varieties of *Papaver somniferum* L. with changed morphine alkaloid. Hereditas 84:69

Ono M, Shimamiñe M, Takahashi K (1977) Gas-chromatographic determination of the major alkaloids in opium. Eisei Shikensho Hokoku 95:4–9

Ono M, Shimamine M, Takahashi K (1978) Gas and high-speed liquid chromatographic determination of the other alkaloids in opiate preparations. Eisei Shikensho Hokoku 96:63–66

Pettitt BC Jr, Damon CE (1982) Analysis of poppy straw concentrate by high-performance liquid chromatography. J Chromatogr 242:189–192

Prista LVN, Fontes DM, Alves da Silva (1976) Chromatographic separation and determination of morphine in opium extracts. An Fac Farm Univ Fed Pernambuco 15:69–76

Šantavý F (1970) Papaveraceae alkaloids. In: Manske RHF (ed) The alkaloids: chemistry and physiology, vol 12. Academic Press, New York, pp 333–454

Šantavý F (1979) Papaveraceae alkaloids II. In: Manske RHF, Rodrigo R (eds) The alkaloids, vol 17. Academic Press, New York, pp 385–545

Shafiee A, Lalezari I, Nasseri-Nauri P, Asgharian R (1975) Alkaloids of *Papaver orientale* and *Papaver pseudo-orientale*. J Pharm Sci 64:1570–1572

Sharma JR, Singh OP (1983) Genetics and genetic improvement. In: Husain A, Sharma JR (eds) The opium poppy. CIMAP, Lucknow (India), pp 39–67

Sharma JR, Lal RK, Mishra HO, Sharma S (1988) Heterosis and gene action for important traits in opium poppy (*Papaver somniferum* L.). Indian J Genet 48:261–266

Srivastava RK, Sharma JR (1987) Estimation of genetic variance and allied parameters through biparental mating in opium poppy (*Papaver somnierum* L.). Aust J Agric Res 38:1047–1052

Srivastava VK, Maheshwari ML (1985) Liquid chromatographic determination of major alkaloids in gum opium. J Assoc Off Anal Chem 68:801–803

Srivastava VK, Pareek SK, Maheshwari ML (1989) Yield and alkaloids profile in poppy capsules. Indian J Pharm Sci 51:133–137

Staba EJ, Zito S, Amin M (1982) Alkaloid production from *Papaver* tissue cultures. J Nat Prod 45:256–262

Szántay CS, Blasko G, Barczai-Beke M, Dorneyi G, Pechy P (1983) Studies on the synthesis of morphine. VI. Planta Med 48:207–211

Thakur RS (1983) Opium: chemistry and uses. In: Husain A, Sharma JR (eds) The opium poppy. CIMAP, Lucknow (India), pp 117–133

Theuns HG, Janssen RHAM, Salemink CA (1987) The alkaloids of the *Papaver* section *Oxytona* Bernb. herbs, spices and medicinal plants. Recent Adv Bot Hort Pharmocol 2:57–110

Vaidya PV, Pundlik MD, Meghal SK (1980) Rapid method for extraction and colorimetric estimation of morphine in Indian opium. J Assoc Off Anal Chem 63:685–688

Verma RK, Uniyal GC, Gupta MM (1990) High-performance liquid chromatography of poppy straw. Indian J Pharma Sci 52:276–278

Vincent PG, Engelke BF (1979) High-pressure liquid chromatographic determination of the five major alkaloids in *Papaver somniferum* L. and thebaine in *Papaver bracteatum* Lindl. capsular tissue. J Assoc Off Anal Chem 62:310–314

Wu Cy, Wittick JJ (1977) Separation of five major alkaloids in gum opium and quantitation of morphine, codeine and thebaine by isocratic reverse phase high-performance liquid chromatography. Anal Chem 49:359–363

Ziegler HW, Beasley TH, Smith DW (1975) Simultaneous assay for six alkaloids in opium using high-performance liquid chromatography. J Assoc Off Anal Chem 58:888–897

Subject Index

MIX
Papier aus verantwortungsvollen Quellen
Paper from responsible sources
FSC® C105338